Shape Makers

Developing Geometric Reasoning

with

THE
GEOMETER'S SKETCHPAD®

Michael T. Battista
Kent State University

KEY CURRICULUM PRESS
Innovators in Mathematics Education

Editors	Dan Bennett and Casey FitzSimons
Geometer's Sketchpad Software Design	Nicholas Jackiw
Editorial Assistant	Jeff Gammon
Teacher Reviewers	Linda Hallenbeck, East Woods School, Hudson, OH
	William F. Halloran, Alternative School, Linden, NJ
	Rosemarie McCabe, Alternative School, Linden, NJ
	Lee Tempkin, Berkwood-Hedge School, Berkeley, CA
Production Editor	Jason Luz
Production Consultant	Steve Rogers
Production Coordinator	Diana Krevsky
Copy Editor	Mary Roybal
Cover Design and Art	Kirk Mills
Interior Design	Christy Butterfield Design
Layout	Ann Rothenbuhler
Publisher	Steven Rasmussen
Editorial Director	John Bergez

Partial support for the development of ideas contained in *Shape Makers* was provided by the National Science Foundation under grant RED 8954664. The views expressed in the book, however, are those of the author and do not necessarily reflect the views of that foundation.

Key Curriculum Press
P.O. Box 2304
Berkeley, CA 94702
510-548-2304
editorial@keypress.com
http://www.keypress.com

Printed in the United States of America   10 9 8 7 6 5 4 3 2 1   01 00 99 98 97   ISBN 1-55953-291-2

This book is dedicated to my wife, Kathy, and children, Jonathan and Emily

ACKNOWLEDGMENTS

Linda Hallenbeck taught several versions of *Shape Makers* as it was being developed. I thank her and her students for letting me observe and talk to them during and after the classroom activities. I also thank Linda for the many helpful comments she has made about the activities. But most of all, I thank Linda for being such a wonderful teacher and colleague. Having her implement the unit has made me feel like a composer whose musical piece is being performed by a world-class orchestra.

I would also like to thank Kathy Battista, Caroline Borrow, and Judy Arnoff for the helpful comments they have made during the writing of this book.

Michael T. Battista

CONTENTS

PHILOSOPHY AND
INSTRUCTIONAL OVERVIEW

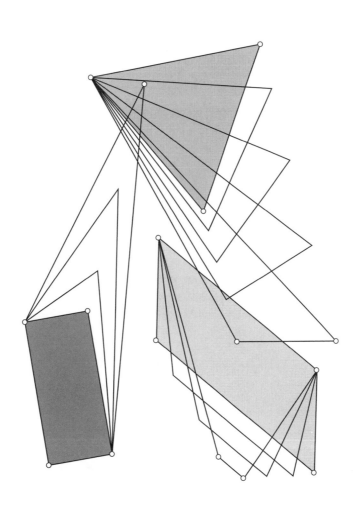

BASIC TENETS OF THE INSTRUCTIONAL APPROACH

Mathematics as Reasoning

Fundamental to the approach taken in this book is my belief that mathematics is first and foremost a form of reasoning, not the performance of endless sequences of procedures invented by others. We do mathematics in our mind, not with our hands or with tools (although *thinking* about what we do with tools can aid tremendously in mathematical thought). To do mathematics is to think in a logical manner; to formulate and test conjectures; to form conclusions, judgments, and inferences. We do mathematics when we recognize patterns, form and manipulate concepts, build and test arguments, invent procedures to solve classes of problems, and decide when to apply procedures we have learned. We do mathematics when we solve problems that have genuine meaning for us. In doing mathematics, we purposefully create, inspect, and manipulate ideas and images to solve problems dealing with quantitative and spatial situations. We reflect on what we know and reorganize it; we make sense of things; we meaningfully and purposefully manipulate mathematical symbols to support our mathematical thought.

Geometry as the Study of Structure

Geometry is the study of ways of organizing or *structuring* our spatial environment and investigating the nature and consequences of that structuring. When we structure something, we determine its nature, shape, or organization by establishing interrelationships between its parts. We structure the plane and space when we organize them by coordinate systems. We structure our visual world when we view it in terms of shapes such as lines, angles, polygons, polyhedra, and geometric transformations.

In this book, students investigate two of the primary classes of shapes used to structure our spatial environment—quadrilaterals and triangles. They examine not only these shapes and relationships between them, but parts of the shapes (such as angles and sides) and

interrelationships between these parts. They also develop and refine their geometric and spatial reasoning and problem-solving skills.

Learning Mathematics in a Culture of Inquiry and Sense Making

Real comprehension of a notion or a theory implies the re-invention of this theory by the subject. Each time one prematurely teaches a child something he could have discovered himself the child is kept from inventing it and consequently from understanding it completely. Naturally, this does not mean that the teacher has no role anymore, but that his role is less that of a person who gives "lessons" and is rather that of someone who organizes situations that will give rise to curiosity and solution-seeking in the child, and who will support such behavior by means of appropriate arrangements. (Piaget)

Learning

Students do not learn by receiving or absorbing ready-made ideas from objects or people. Instead, they learn as they reflect on and abstract the mental and physical actions they perform while purposefully interacting with their physical and social environments. As they interact with these environments, they construct mental structures that enable them to make sense of and manage their physical, social, and intellectual experiences.

Like scientists, students are theory builders. They learn as they reorganize their theories, as they discover and adopt more sophisticated and general theories. Such reorganization is triggered by perturbations—that is, by students' realization that their current way of interpreting things does not work or produces unexpected results.

Teaching

The goal of instruction should be to help *each* student build mathematical ideas and theories that are more complex, abstract, and powerful than those he or she currently possesses. The major instructional mechanism for encouraging students' construction of knowledge is the presentation of properly chosen problematic tasks. These tasks guide the direction of students' theory building by properly focusing their attention, encouraging them to reflect on their actions and thoughts, and promoting perturbations that require reorganization of current theories.

However, to be effective, instructional tasks must fall within the students' current *zones of construction*. That is, students' construction of the new concepts required to complete the tasks must be possible given their current conceptual structures and operations. In fact, because students' existing structures determine how they think about all new tasks, as teachers, we must constantly monitor the development of these structures and adjust our instruction accordingly.

Whenever we ignore students' current ways of thinking and attempt to impose methods on students, the sense-making activity of students is stifled. Students mindlessly mimic the methods they are shown. Their belief about the nature of mathematics changes from seeing mathematics as sense-making to seeing mathematics as the learning of set procedures that make little sense. Students change from intellectually autonomous thinkers to teacher/textbook-dependent rule followers.

Establishing a Culture of Inquiry

A fundamental tenet in current research-based, scientific theories of learning mathematics is that instruction should be inquiry-based, with students learning mathematics as they solve problems and share their ideas with one another. To foster meaningful learning in the classroom, teachers, in collaboration with their students, must establish a culture of inquiry in which individuals pose questions, solve problems, share ideas, and think critically. Within this culture, students are involved not only in inquiry, problem solving, and invention, but also in classroom discourse that establishes ideas and truths collaboratively. Students' participation in such a culture promotes their personal construction of ideas as they

- attempt to elaborate and clarify personally developed ideas so that they can communicate them to others;

- reflect on, evaluate, and justify their personally developed ideas in response to challenges posed by classmates;
- attempt to make sense of and sometimes utilize new ideas offered by classmates.

Responsibilities in a Culture of Inquiry

The major responsibilities of teachers and students in a classroom culture of inquiry are as follows.

Students are responsible for:

1. Attempting to solve and make sense of all problems given to them.
2. Explaining their mathematical thinking to other members of the class and justifying problem solutions in response to challenges.
3. Listening to, as well as attempting to make sense of, other students' mathematical explanations and problem solutions. This includes
   - asking for clarification if an explanation is not understood;
   - challenging strategies and problem solutions that do not seem reasonable.
4. Working collaboratively with other students. This includes attempting to reach consensus on problem solutions while respecting the rights of others to derive or justify solutions differently.

Teachers are responsible for:

1. Selecting instructional tasks and guiding students' work on these tasks so that students' thinking becomes increasingly more sophisticated. This includes
   - choosing sequences of problematic tasks that are based on detailed knowledge of how students construct meanings for specific mathematical topics as well as the conceptual advances that students can make with those topics during the course of instruction;
   - continually assessing students' learning progress and adjusting instruction accordingly;
   - encouraging students to reflect on their mathematical experiences.

2. Establishing a social environment that supports a spirit of inquiry and collaborative small-group work. This includes

- explaining;
- illustrating with classroom examples;
- regularly reminding students of their responsibilities (as described earlier).

3. Encouraging productive dialogue among students. This includes

- encouraging students to explain and justify their mathematical ideas;
- highlighting conflicts between alternative student interpretations or solutions;
- unobtrusively encouraging potentially fruitful student contributions;
- redescribing students' ideas in more sophisticated ways that students can still comprehend;
- introducing mathematical concepts, symbolism, and terminology at appropriate times so that students can use them to reflect on and communicate about their own developing ideas.

Helping Students Meet Their Responsibilities.

If students are not accustomed to participating in a culture of inquiry in mathematics class, it may take them several weeks to become comfortable with and competent within such an instructional environment. They will need regular and explicit discussion of this new way of learning. Posting student responsibilities on a poster board in class and regularly asking students what their responsibilities are and how they are implementing them can help.

Students will also need to see classroom examples of their responsibilities in action. For instance, to learn how to explain and justify their strategies and problem solutions, students need to talk about the processes of explanation and justification. If you see that a number of students are having difficulty explaining their thinking about a particular problem, make such explanations the focus of a class discussion. You might start the discussion by asking students who are having difficulty articulating an idea to explain it as best they can to the class. Then ask other students who used a similar strategy how they explained it: "How do you think we should talk about these ideas? What words should we

use to refer to what you are talking about?" Such discussions can help students develop a language and set of conceptualizations for describing their developing ideas.

Understanding Students' Geometric Thinking

A considerable amount of research has established the van Hiele theory as an accurate description of the development of students' geometric thinking. Knowledge of these levels is essential in designing, conducting, and evaluating meaningful geometry instruction.

The van Hiele Levels

According to van Hiele, students progress through several levels of qualitatively different and increasingly sophisticated levels of thought in geometry.[1] They pass through these levels sequentially. Consequently, students who are required by instruction to study content at a higher level than they have achieved cannot make sense of that content, and they resort to memorization. Furthermore, people who reason at different levels may use the same terms but have very different meanings for those terms. Thus, effective communication between people at different levels—especially teacher and students—can be difficult.

In the van Hiele theory, a critical factor used in distinguishing levels of thinking is how students deal with *geometric properties*. Such properties describe spatial relationships between parts of shapes. For example, these statements describe geometric properties:

a. Opposite sides of this quadrilateral are congruent.

b. Opposite sides of parallelograms are parallel.

c. A rectangle has four right angles.

d. Adjacent angles of parallelograms are supplementary.

e. This quadrilateral has one line of symmetry.

[1] For more details see D. H. Clements and M. T. Battista, "Geometry and Spatial Reasoning" in *Handbook of Research on Mathematics Teaching and Learning*, ed. D. Grouws (New York: NCTM/Macmillan, 1992), 420–464.

Examples (a) and (b) explicitly describe relationships between the sides, or parts, of quadrilaterals. By specifying the measures of angles, examples (c) and (d) describe relationships between pairs of adjacent sides. Finally, example (e) describes a relationship between two "halves" of a quadrilateral, which, with further analysis, could be described in terms of line segments and angles.

Level 1: Visual. At the first van Hiele level, students identify and reason about shapes and other geometric configurations according to their appearance. Their thinking is dominated by perception. They recognize and mentally represent shapes such as squares and triangles as visual wholes. When identifying shapes, students often use visual prototypes, saying that a given figure is a rectangle, for instance, because "it looks like a door." Students at the visual level do not attend to geometric properties of shapes. For example, they might distinguish one shape from another without referring to a single property of either shape. Instead, they might judge that two shapes are congruent because they look the same or because they can be turned to look the same.

Level 2: Descriptive/Analytic. At the second van Hiele level, students recognize and can characterize shapes by their properties, that is, by spatial relationships between their parts. For instance, students might think of a rectangle as a figure that has opposite sides equal and parallel as well as having four right angles. While still important, the appearance of shapes becomes secondary because students conceptualize shapes as being determined by collections of properties rather than as simply matching visual prototypes. Properties are established experimentally by observing, measuring, drawing, and model making. However, students tend to name all the properties they know for a class of shapes, rather than a sufficient set. They also do not see relationships between classes of shapes (e.g., a student might contend that a figure is not a rectangle because it is a square).

Of course, which properties students attribute to shapes depends on their experiences with those shapes. For instance, although students typically come to see that rectangles have right angles and congruent and parallel opposite sides,

it is not likely that they will notice that the diagonals bisect each other unless they have had sufficient experience analyzing diagonals. Furthermore, some students formulate incorrect properties of shapes. For example, many middle-school students think that rectangles cannot have all sides congruent.

Level 3: Abstract/Relational. At the third van Hiele level, students can form abstract definitions, distinguish between necessary and sufficient sets of conditions for a class of shapes, and understand and sometimes even provide logical arguments in the geometric domain. They can classify shapes hierarchically and give informal arguments to justify their classifications (e.g., a square is identified as a rhombus because it can be thought of as a "rhombus with some extra properties"). They can understand why, and are willing to accept that, a square is a rectangle. They can discover properties of classes of figures by informal deduction. For example, they might deduce that in any quadrilateral the sum of the angles must be 360° because any quadrilateral can be decomposed into two triangles, each of whose angles sum to 180°. However, for students at this level, "any trapezoid" may actually mean "all the trapezoids with which I am familiar," not necessarily all possible trapezoids.

Because students see that some properties imply others, they no longer feel a need to list all the properties of a class of shapes. Definitions are seen not merely as descriptions of shapes but as a way of logically organizing properties. The students still, however, do not grasp that logical deduction is the method for establishing geometric truths.

Level 4: Formal Deduction and Proof. At the fourth van Hiele level, students can formally prove theorems within an axiomatic system. That is, they can produce a sequence of statements that logically justifies a conclusion as a consequence of the "givens." They recognize the difference among undefined terms, definitions, axioms, and theorems. They can reason by employing formal logic to interpret geometric statements.

Thinking at level 4 is required for a proof-oriented high school geometry course. Of course, an even higher level of thought is needed to analyze and compare different axiomatic systems.

Transitions Between Levels. During the transition from the visual to the descriptive/analytic level, students start attending to the components of shapes such as sides and angles. They start examining the relationships between these components. But often their descriptions of these relationships are intuitive, visual, and imprecise. For instance, students might say that the difference between a rectangle and a parallelogram with no right angles is that the sides of the parallelogram are tilted and the sides of the rectangle are straight.

In the transition from the descriptive/analytic to the abstract/relational level, students begin to discover that some combinations of properties of a class of shapes imply other properties. For instance, they might claim that because a rectangle has opposite sides parallel, the opposite sides must be equal.

At What Levels Are Middle-Grade Students Functioning?

Most middle-grade and junior-high students are functioning at van Hiele level 1 or 2.[2] In fact, over 70% of students begin high school geometry at level 1 or below. Unfortunately, research indicates that only students who enter at level 2 or higher have a good chance of becoming competent with proof—a level 4 activity—by the end of the course. Because so many students are functioning at such low van Hiele levels when they enter high school geometry, such courses are dismally ineffective. Indeed, almost 40% of students end the year below level 2, and only about 30% in courses that teach proof reach a 75% mastery level in proof writing.

Thus, a major goal for middle-grade and junior-high geometry curricula—and that of the *Shape Makers* curriculum—should be to help students move from level 1 to level 2, then to level 3 in the van Hiele hierarchy. Students who have worked through the *Shape Makers* curriculum will be well prepared for high school geometry courses that start with further development and expansion of students' level 3 reasoning before dealing with formal proof.

[2] See Clements and Battista, 1992, for more details on this research.

THE *SHAPE MAKERS* COMPUTER MICROWORLD

The usual approach to teaching students about classes of geometric shapes is to define the shapes. However, as was discussed earlier, only students at van Hiele level 3 can fully understand definitions such as "A rectangle is a quadrilateral with four right angles" or "A square is a rectangle with all sides equal." So the definitional approach is too formal to have much chance of success with most students. It forces students to attempt to find their way in what, for them, is an incomprehensible maze of meaningless abstractions. It is little wonder that so few students succeed, and that fewer still enjoy what they are doing.

Shape Makers provides an alternative approach. Students can manipulate the Shape Makers just as they can manipulate a physical apparatus. Through their actions and reflection on those actions, students can discover properties of the Shape Makers, that coincide with those of the classes of shapes made by the Shape Makers. In essence, students can learn about properties and classes of shapes using the same processes they use in learning everyday concepts like "chair" or "book." That is, they can manipulate and reflect on numerous examples instead of trying to comprehend verbal definitions. Eventually, after extensive visual investigations have enabled students to understand shapes in terms of their properties, students can deal meaningfully with geometric definitions.

Dynamic Mental Models

The *Shape Makers* microworld is designed to promote the development of dynamic mental models for thinking about geometric shapes and their manipulation. Mental models are mentally constructed representations of real-world situations.

They are derived from our experiences and our reflections on those experiences, and they usually have an image-like quality. Reasoning with mental models is like reasoning about physical objects. When using a mental model to reason about a situation, a person can mentally move around, move on or into, combine, and transform objects, as well as perform other operations like those that can be performed on objects in the physical world. Students draw inferences by mentally manipulating mental models and observing the results.

The primary source of mental models is our experience in dealing with the world, especially with physical objects. To think of how a mental model for a parallelogram might be derived from real-world manipulation, imagine four straight rods connected at their endpoints in a way that permits freedom of movement at the connections—a movable quadrilateral. Imagine now that the opposite rods are the same length. No matter how we move this physical apparatus, it always forms a parallelogram, and sometimes a rectangle. See Figure 1 below.

As we manipulate our "parallelogram maker," we not only see how its shape changes, we feel the physical constraints that we have built into it. We see and feel how one parallelogram is related to others. The visual and kinesthetic experiences that we abstract from our actions with this apparatus, along with our reflections on those actions, are integrated to form a mental model for a parallelogram, a model that we can use in reasoning about parallelograms.

The power of utilizing mental models to reason about geometry can be illustrated by the case of a second-grader who had been contemplating the notion that squares are special types of rectangles.

Figure 1

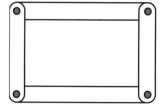 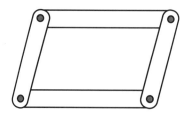 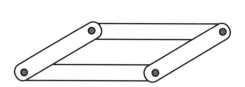

She made sense of the idea not by referring to verbally stated properties of these shapes, but by thinking about how some "stretchy square bathroom things" could be stretched into rectangles.[3] She reasoned by performing a simulation of changing a square into a rectangle using the mental model she had derived from her physical actions with the "stretchy square bathroom things." Furthermore, she performed a special type of visual transformation, one that incorporated some formal mathematical constraints—preserving 90° angles—into her mental model. Her reasoning was intuitively constrained by her emerging knowledge of the properties of shapes. Such reasoning is the key to meaningful and powerful geometric thinking.

The *Shape Makers* Computer Microworld

The *Shape Makers* computer microworld is built upon The Geometer's Sketchpad®, a software tool for constructing and investigating geometry dynamically. The *Shape Makers* microworld is designed to help students construct appropriate mental models for thinking about various types of quadrilaterals and triangles. In this microworld, each class of common quadrilaterals and triangles has a Shape Maker, a Sketchpad™ construction that can be dynamically transformed in various ways, but only to produce different shapes in the same class. Not only can these Shape Makers be manipulated like the physical parallelogram maker described above, but their side lengths can also be changed. The Rectangle Maker, for instance, can be manipulated to make any desired rectangle that fits on the computer screen, no matter what its shape, size, or orientation—but it can make only rectangles. See Figure 2 at right.

The Rectangle Maker's shape is changed by using the mouse to drag one of its control points. A *control point* is represented by a small circle at one of the vertices of the Rectangle Maker. To drag a control point from one location to another, point to it with the arrow, press and

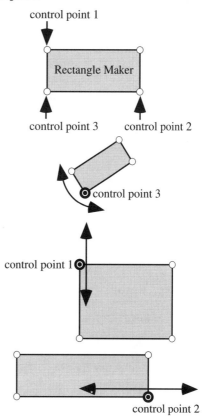

Figure 2

control point 1

Rectangle Maker

control point 3   control point 2

control point 3

control point 1

control point 2

hold down the mouse button (the point becomes highlighted, indicating that it is selected), then move the mouse, keeping the button held down. Release the mouse button when the control point is at the desired location.

The Rectangle Maker can be made taller or shorter with control point 1 and wider or narrower with control point 2. Its position and size can be changed with control point 3. (Although the exact functions of the different control points vary among Shape Makers, you can always change a Shape Maker's size and position, along with all of its critical attributes, such as side lengths and angle measures.)

You can move the Rectangle Maker from one screen location to another by dragging its interior. When the Rectangle Maker is in the desired position, click somewhere off the Rectangle Maker to deselect it.

There are quadrilateral Shape Makers for squares, rectangles, parallelograms, kites, rhombuses, trapezoids, and general quadrilaterals. There are triangle Shape Makers for general, isosceles, equilateral, and right triangles.

[3] Battista, M. T. "On Greeno's Environmental/Model View of Conceptual Domains: A Spatial/Geometric Perspective" in *Journal for Research in Mathematics Education* (January 1994) 25: 86–94.

It is important to recognize that the *Shape Makers* computer microworld has been constructed in The Geometer's Sketchpad, which is a comprehensive, dynamic geometry construction program. In Sketchpad, students can construct, measure, and transform geometric shapes on the computer screen. Measurements are instantaneously updated as geometric objects are altered. For instance, if you construct and measure the line segment between two points, then move one of the points, the lengths of the newly created segments are displayed continuously. Furthermore, geometric constraints that are built into constructions can be preserved as the shapes are varied. For instance, if you construct a polygon and reflect it about a line, then move one of the polygon's vertices, the resulting changes to the reflected image are made automatically and instantaneously.

Students who are working with the Shape Makers need to know very little about using Sketchpad—what they do need to know is explained within the activities. In fact, the tools and some menus have been hidden from view in the Shape Maker sketches so that students don't inadvertently activate tools they don't need in the activities. The only tool students need is the Selection Arrow tool, simply called the arrow here. However, there are times when you or your students might enjoy extending *Shape Makers* analyses and activities using commands in Sketchpad. To become familiar with the possibilities, see *The Geometer's Sketchpad User Guide and Reference Manual.* One of the advantages of using the Shape Makers is that students start to become familiar with a computer tool—The Geometer's Sketchpad—that they can use productively for the rest of their mathematical careers.

The *Shape Makers* Disks

The two disks that accompany this book each contain an identical collection of Sketchpad *Shape Makers* sketches. One disk is for Macintosh® computers and the other is for computers running Microsoft® Windows® 3.1 or later. The sketches on the Macintosh disk are compressed into a single self-extracting archive in order to fit on a single floppy disk. To use them, you'll need to extract them to a computer's hard disk by double-clicking the archive icon:

Shape Makers Sketches

When you do, you'll be asked to select a destination folder. As a Sketchpad user, you should have a folder titled The Geometer's Sketchpad on your hard disk. Open it or any other location on your hard disk where you wish to save the *Shape Makers* sketches. Click **Extract**. All the sketches will be copied into a single folder on your hard drive called *Shape Makers*. Use the disk to extract the sketches to as many computers in your classroom or lab as you like, or extract them to your network file server.

The Windows sketches are not compressed. They can be run off the floppy disk, but they'll work better if they're copied onto a hard disk.

Accessing Shape Maker Files

Shape Makers are contained in special Sketchpad files called sketches. Once the Sketchpad program and Shape Maker sketches have been loaded onto a computer's hard disk, the sketches can be accessed by choosing **Open** from Sketchpad's **File** menu. In the Open dialog box that then appears, highlight and open the appropriate folder or directory; then highlight and open the Shape Maker sketch or sketches for that particular activity. The organization and location of the sketches on the Shape Makers disk are shown in the table on the following page. (Folders are preceded by a ☐ icon; Shape Maker sketches are preceded by a ⬙ icon.)

In the description of the instructional activities, the sketches needed for each exploration are listed just before the "Required Materials" chart. For example, the three Shape Maker sketches needed for Quadrilateral Exploration 1 are displayed like Figure 3 on the following page.

On student sheets, sketches are listed the same way, right after the title of the activity.

Important Note. Any time the computer asks students if they want to save their work, they must choose Don't Save. Otherwise, the changes will be made to the original Shape Maker files on the hard disk.

File Organization Table

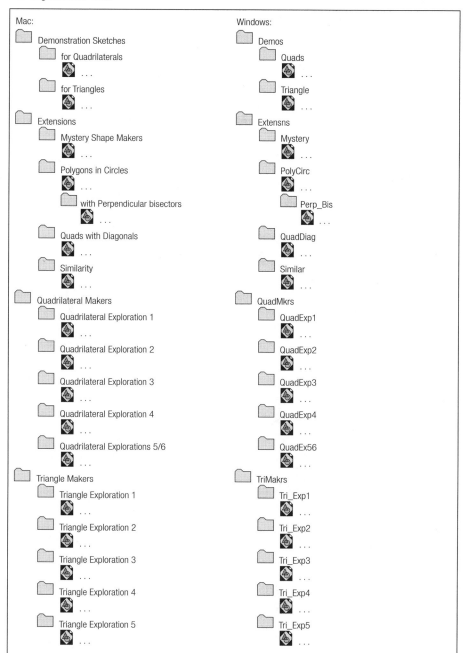

Figure 3

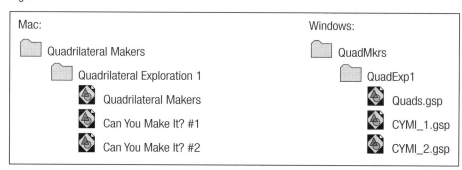

INSTRUCTIONAL OVERVIEW

Goals

The set of activities in this book is intended to replace the standard textbook treatment of geometry in the middle and junior-high grades. It can also be used to replace early portions of the high school geometry curriculum. However, while these activities cover the major ideas dealt with in standard curricula, they do not progress through each idea in a lockstep manner. It is intended, instead, that students meaningfully construct new concepts and ways of reasoning as they solve interesting problems in a culture of inquiry.

The major goals for the instructional activities in this book are for students to

- learn to identify different types of quadrilaterals and triangles, both visually and by analyzing the shapes' measurements;
- learn the properties of different classes of quadrilaterals and triangles;
- learn how to define and classify quadrilaterals and triangles;
- learn about the different ways that triangles can be rotated and reflected to make quadrilaterals;
- learn to make, as well as justify, conjectures and inferences in plane geometry;
- learn to reason, communicate, make conceptual connections, and solve problems in spatial and geometric contexts;
- move from level 1 to level 2, then into the partial attainment of level 3, in the van Hiele hierarchy of geometric thinking.

Classroom Setup

The activities in this book should be completed by pairs of students working collaboratively. Each student gets a student sheet for each activity. Students talk about solutions to problems, but they write their own answers on their student sheets.

Students in each pair should stay together for the whole unit (although if you split the unit between quadrilaterals and triangles, you could change

pairs then). Keeping students in the same pairs enables them to build up a history of collaboration, concept building, and communication, all of which are necessary for cooperatively constructing new ideas. Switching pairs inhibits students' natural tendency to constantly build on previously formulated ideas.

To enhance the productivity of small-group learning, pair students based on your knowledge about (a) how they will get along and interact with each other and (b) their ability levels in mathematics. As a general rule, if you could divide your students into three ability levels in mathematics—low, middle, and high—forming a pair out of two students in the same level or two adjacent levels (low/middle, middle/high) works well, while pairing a low and a high student generally does not.

Each class session lasts about one hour. Class discussion of students' work comes after all students have completed a student sheet. These discussions are moderated, guided, and facilitated by you, the teacher, but students make almost all of the mathematical contributions. The discussions, therefore, are an essential part of establishing and maintaining a culture of inquiry. If your class periods are shorter than 60 minutes, be sure to allow ample time for follow-up discussions during the next period.

Each student should keep a journal of thoughts and activities in the unit. Journals should include completed student sheets and any thoughts students had while completing the activities or participating in class discussions.

Computer Management

Most of what students will be doing on the computer requires only the use of the mouse to choose menu commands and to manipulate the Shape Makers. No keyboarding skills are required.

Ideally, there should be one computer for each pair of students. Establish rules that specify how partners should take turns being in control of the mouse. For instance, students can alternate each day at the midway point of the period or every other day. Within this alternating scheme however, students should be reasonable about allowing their partners to use the mouse whenever their partner has a good idea or is having difficulty that can be resolved only by personally manipulating a Shape Maker.

Whenever students (or adults!) are sitting at a computer screen, it is difficult to hold their attention in a discussion. Thus, you need to take specific steps to ensure that students are attentive whenever you are giving directions or making comments they need to hear. The following procedures have worked well for teachers using computers.

Directions about the use of the computer and *Shape Makers* software should be given using a single computer, with its screen projected so that all students in the class can see it or even with all students gathered around a single computer screen. This keeps students' attention on what you are saying, especially if you ask plenty of questions. Keep in mind, though, that some students will not understand exactly what to do until they try out what you have said at their own computers. So you will need to help students with some operations after they go to their computers.

Class discussions are best held with students away from their computers and in possession of their student sheets and notes. Ideally, there will be enough space in your classroom for the computers and for a discussion area so that transitions between group work and class discussions can be made smoothly.

Occasionally, you may need to talk to students while they are working in groups at the computers. If what you need to say is not brief, you should have students move away from the computers into the discussion area. If it is brief, you need to ensure that students don't get distracted from your comments by the lure of the computer. Many teachers specifically tell students to put their hands on their knees or in their laps, for instance, to prevent them from continuing their computer work.

Finally, some teachers find it useful to have some type of signal at each computer station—usually on the monitor—that indicates the status of students' work. For instance, a card can be folded into a triangular prism, with each of its three lateral faces a different color. On one side, the word HELP can be written, indicating that students need help. On another side, the word DONE can be written, indicating that students have completed the assigned task (and have gone on to another task that you have specified for those who have completed the initial task). Finally, the word WORKING could be written on the third side, indicating that students are working on the initial task that was assigned. By looking at these signals, you will be able to quickly locate students who need help (without students constantly getting out of their seats to talk to you). You will also be better able to gauge how many students have completed the assigned task.

The Importance of Student Predictions

As discussed earlier, reflection is an essential part of learning mathematics. One way to encourage reflection is to have students make predictions, then check their predictions on the computer with Shape Makers. This is done consistently in *Shape Makers* activities. Students should constantly reflect on discrepancies between their predicted answers and those found using the Shape Makers. Making predictions before using the Shape Makers encourages students to form and organize mental models and theories about shapes. As they check their predictions, they refine these models and theories.

Talk to students about making predictions. They shouldn't feel bad if their predictions are incorrect; the activities will help them improve their understanding. Tell students that making and thinking about predictions will help them develop increasingly powerful reasoning about shapes, whether their predictions are correct or incorrect.

Cultivating Conjectures

Throughout the book, encourage students to write their conjectures and ideas on the provided Conjectures and Queries student sheets (which can then become part of their journals). You should make copies of this student sheet available to students, and discuss ideas periodically. Students can make conjectures or queries about the properties of specific Shape Makers or classes of shapes. They can make conjectures or queries about relationships between different Shape Makers or between different classes of shapes.

To promote reflection on these conjectures and queries, also have students write their entries on a classroom chart. Periodically, you might summarize these conjectures and queries and distribute them to the whole class. An ongoing task, which can be worked on after students have completed assigned Explorations, is to reflect on conjectures generated by other students, trying to confirm or disconfirm them.

Justifying Ideas

In inquiry-based classrooms, primary responsibility for establishing the validity or "truth" of mathematical ideas lies with students, not with teachers or textbooks. In essence, each student is seen as a mathematician, somebody who is responsible for solving mathematical problems, making conjectures, and establishing the validity of those conjectures within the classroom culture. The teacher does not act as the mathematical authority. In fact, *whenever* students make statements that describe their mathematical problem-solving activity, they are implicitly claiming that such statements are valid. Thus, their explanations should be elaborate enough for other students to evaluate their validity.

To help students' justification of mathematical ideas gradually develop and become increasingly rigorous, instruction should focus on students' building arguments to convince themselves and others of the validity of their ideas. It should permit students to utilize visual and empirical thinking, because such thinking is the foundation for higher levels of geometric reasoning. It should involve students in the crucial elements of mathematical discovery and discourse—explanation, conjecturing, careful reasoning, and the building of justifying arguments that can be scrutinized by others. In this atmosphere, students will eventually see the limitations of visual and empirical attempts at validation and will move toward logical, deductive methods.

Assessment

Assessment is an ongoing activity that involves daily observation of students, as well as inspection of appropriate samples of students' individual written work. No matter what the format, the goal of assessment should be to reveal the progress each student is making in attaining increasingly sophisticated geometric reasoning as described in the section "Understanding Students Geometric Thinking" in this introduction.

Specific suggestions for assessing particular topics are given throughout the book. However, one overall assessment strategy that teachers have found useful is to have students keep all their student sheets in Shape Maker folders. If you have students insert additional sheets that contain their thoughts about specific ideas—as if in a journal—these folders become portfolios of the students' work. These portfolios can then be used to analyze the development of students' geometric reasoning as they work through *Shape Makers*.

ORGANIZATION OF THE BOOK

General Organization

Shape Makers is organized into *Explorations*, each of which consists of a set of related instructional one-hour class sessions. Each Exploration provides a detailed description of the instructional activities in that Exploration, in addition to listing mathematical objectives, required Sketchpad sketches, required student sheets, and other necessary materials. Many Explorations include Mathematical Notes or Teaching Notes that elaborate important mathematical or pedagogical considerations relevant to that Exploration. Suggestions for homework, extensions, and assessment are also provided. Suggested teacher dialogue and questions are set off in italics. Specific actions that the teacher needs to take during the lesson are preceded by a black arrow: ➡.

Teaching Notes discuss aspects of teaching and usually are illustrated with student and/or teacher dialogues. These dialogues are taken from actual classrooms, but they have been edited and shortened to make them more readable. Nevertheless, some of the dialogues require careful reading, because students' thoughts about the ideas covered in this book are often quite complex. I have included these extensive dialogues because understanding students' thinking is a critical component in effectively using the instructional approach taken in this book. *Mathematical Notes* discuss mathematical issues that require some special attention.

Instruction starts with quadrilaterals, because there are more interesting relationships between types of quadrilaterals than between types of triangles. A nice hierarchical classification exists for quadrilaterals. Activities are sequenced to guide students' thinking about quadrilaterals from visual to property-based, and then to thinking that deductively interrelates classes of shapes.

After students have studied quadrilaterals, types of triangles are introduced. As with quadrilaterals, triangle activities are sequenced to guide students' thinking about triangles from visual to property-based to classificatory reasoning. Finally, relationships between types of triangles and types of quadrilaterals are explored, and the two types of shapes are combined to make tessellations.

Glossary of Geometric Concepts

As a reference, the geometric concepts covered in this book are described in the "Geometric Glossary" on page 147. Avoid giving these descriptions to students. Students should personally construct their own notions of these ideas as they solve and explore problems with the Shape Makers and as they discuss these ideas with one another.

Overview of Explorations

This section gives a brief overview of the content of the instructional activities and how they are interrelated. Although the activities were designed to be used as a complete sequence, this overview provides some guidance about what activities might be omitted if you have limited time or if you plan a more limited coverage of ideas.

Shape Makers attempts to gradually move students from the first van Hiele level (thinking about shapes visually and holistically) to the second (thinking about shapes in terms of their properties) and finally to the third (forming definitions for shapes, thinking about interrelationships between classes of shapes, and moving toward more deductive geometric thought). Generally, Quadrilateral Explorations 1 and 2 and Triangle Exploration 1 focus on firming up students' level 1 thinking. Quadrilateral Explorations 3–5 and Triangle Explorations 2–5 focus on developing level 2 thinking. Quadrilateral Explorations 5 and 6 and Triangle Explorations 4 and 5 focus on level 3 thinking. While students are developing more sophisticated geometric reasoning, they are also learning numerous geometric concepts that support that reasoning, all in the context of solving interesting problems.

As you can see from this overview, it is important for students to progress through the material in the presented sequence, starting from the beginning. However, you can omit some activities without endangering the developmental nature of the instruction. For instance, in both the quadrilateral and the triangle sections, the material becomes more sophisticated as you progress through the section, so the later material in each section can be omitted if your goals do not include moving students to level 3 thinking. You might also omit the material concerning angle relationships when parallel lines are cut by a transversal. Although this will cause students to approach parallelism more intuitively, those intuitions are quite sufficient for the rest of the instructional sequence (except, perhaps, for those portions of the activity "Making Special Quadrilaterals from Triangle Pairs" in which students have to "prove" their answers).

Some activities were developed as fun ways to apply previously encountered ideas. The "Mystery of Polygon Flats" in the Quadrilateral Explorations and "Triangle Tessellations" in the Triangle Explorations are examples. Although these activities can be skipped, students get quite interested in them and, because of this heightened interest, often make significant increases in their understanding of relevant geometric concepts. Another activity that falls in this category is "Shape Maker Riddles." Note, however, that although the concept of symmetry is embedded throughout the instructional activities, the riddles provide an opportunity for students to explore it more explicitly.

QUADRILATERAL EXPLORATIONS

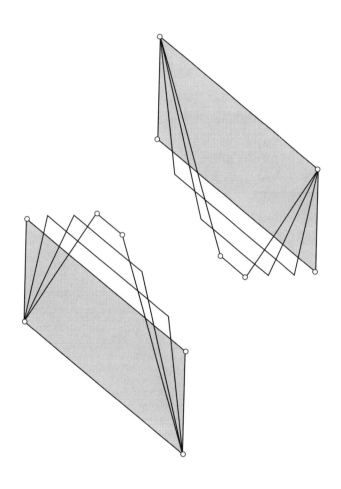

Shape Maker Pictures

Summary

Students use quadrilateral Shape Makers first to make their own pictures, then to copy pictures given to them on student sheets and computer screens.

Mathematical Objectives

Students explore the operation of the different quadrilateral Shape Makers, as well as the kinds of shapes each Shape Maker will make. They start formulating rules that the various Shape Makers follow. These activities help students refine their visual knowledge of various types of quadrilaterals, laying the foundation for their formulation of property-based knowledge of the shapes.

Mac:

📁 Quadrilateral Makers

    📁 Quadrilateral Exploration 1

        📄 Quadrilateral Makers

        📄 Can You Make It? #1

        📄 Can You Make It? #2

Windows:

📁 QuadMkrs

    📁 QuadExp1

        📄 Quads.gsp

        📄 CYMI_1.gsp

        📄 CYMI_2.gsp

Required Materials

Session	Student Sheet	SS#	Geometer's Sketchpad sketch
1	Make Your Own Picture	1	Quadrilateral Makers (Mac) Quads.gsp (Windows)
2–4	Can You Make It? #1	2	Can You Make It? #1 (Mac) CYMI_1.gsp (Windows)
	Can You Make It? #2	3	Can You Make It? #2 (Mac) CYMI_2.gsp (Windows)

Make classroom charts for "Can You Make It? #1 and #2." One way to make a large copy of the student sheet pictures is to use an overhead projector to project them onto large sheets of paper and then trace the projection.

Explaining the Operation of the Shape Makers

Class Discussion

➡ Explain the operation of the Shape Makers to students by using a computer connected to an overhead display or by gathering all students around a single computer.

Do not define the different types of quadrilaterals. Students will do this themselves after much exploration with the Shape Makers. Here is how the Shape Makers should first be introduced to students. (It is essential that students be told that, for instance, the Rectangle Maker was designed to make rectangles and only rectangles.)

The Shape Makers can be used to make different shapes on the computer screen. For instance, the Rectangle Maker can be used to make any desired rectangle that fits on the computer screen, no matter what its shape, size, or orientation— but only rectangles. So any rectangle on the screen can be made by the Rectangle Maker, and all shapes made by the Rectangle Maker are rectangles.

Similarly, any parallelogram on the screen can be made by the Parallelogram Maker, and all shapes made by the Parallelogram Maker are parallelograms, and so on for the other Shape Makers.

➡ Illustrate making rectangles with the Rectangle Maker by dragging its control points, represented by the little circles at its vertices.

I am dragging the Rectangle Maker's control points using the Selection Arrow tool—the little arrow that is moving around on the screen. I move the arrow by moving the mouse. I move a control point by placing the arrow onto it, then holding down the mouse button as I move the mouse.

You can move the Rectangle Maker from one screen location to another by dragging its interior, the light yellow region inside it. To drag the interior, point to it by moving the arrow onto it with the mouse, hold down the mouse button, then move the mouse. When the Shape Maker is in the desired position, release the mouse button, move the arrow to a blank part of the screen, and click once to deselect everything.

Seven types of Shape Makers make quadrilaterals: the Square Maker, Rectangle Maker, Parallelogram Maker, Kite Maker, Rhombus Maker, Trapezoid Maker, and Quadrilateral Maker.

➡ Discuss the use of the Shape Maker control points.

Different control points do different things. [Illustrating with the control points on the Rhombus Maker, ask questions such as:] What does this control point seem to do? How about this one?

What happens with the control points on one Shape Maker may be different from what happens on another Shape Maker. So when you're trying to make a shape with a particular Shape Maker, it's important to try all the control points. As you use the Shape Makers, you will gradually learn to effectively manipulate them with their different control points.

➡ Explain that certain rules must be followed when using the quadrilateral Shape Makers:

Rule 1. Vertices can't overlap.

Rule 2. Sides can't intersect, except at their endpoints.

Rule 3. No three vertices can lie on the same line.

If either of the first two rules is violated, the computer prints a message on the screen to alert students that the figure formed by the Shape Maker is no longer a quadrilateral. Ask students why rule 3 is necessary. (If three vertices lie on the same line, the figure formed will be a triangle, not a quadrilateral.)

➡ Explain opening and closing Shape Maker sketches (as described in "Accessing Shape Maker Files" in the introduction).

Whenever students want to quit using a Shape Maker sketch, they should click on the Close box of the sketch window and be sure not to save their work.

Activity: Make Your Own Picture

Use Student Sheet 1, and sketches:

Mac:

📁 Quadrilateral Exploration 1

◈ Quadrilateral Makers

Windows:

📁 QuadExp1

◈ Quads.gsp

Students Work in Pairs

➡ Distribute the student sheet "Make Your Own Picture" to the class.

Students are to make a picture on the screen using all seven of the quadrilateral Shape Makers. They should draw their final picture on their student sheet, label each part of the picture with the name of the Shape Maker that was used to make it, then answer the questions on the sheet. When they are done, they may make another picture.

➡ Walk around the room and help students who are having technical difficulties. Be sure they have opened the correct sketch and are able to manipulate the Shape Makers.

➡ Ask questions to see how students are thinking about the use of the Shape Makers.

Why did you use the Parallelogram Maker to make this shape? Could you have used another Shape Maker?

➡ Tell students that, throughout the unit, they should write any interesting observations, conjectures, and questions about their work with Shape Makers on a Conjectures and Queries sheet (see page 220). Place an ample supply of copies of this sheet in a convenient location in the classroom.

➡ Also encourage students to write their conjectures and queries on a classroom Conjectures and Queries chart.

Whenever they finish an Exploration early, students should examine the chart and investigate other students' ideas. Periodically, you can summarize what is on the classroom chart and distribute copies of the summary to the students.

Conjectures and Queries

Emphasize to students that doing mathematics consists of inquiry—posing problems, devising and testing possible solutions to problems—not just finding answers. For example, one teacher emphasized this idea in this way:

Teacher: Lots of people disagree about whether the Trapezoid Maker can make this shape. So that's something you all will have to check out.

Brenda: I think it might work, it's my theory—I'm not really sure about it.

Teacher: Are we done with this?

Students: No.

Teacher: We're only getting started. So it's great to come up with *conjectures*— things you think might be true—and *queries*—questions that you would like to answer. It's okay to be puzzled, because we'll try to solve your puzzles as we continue working.

Waiting for the Need

After students have gained enough experience with the Shape Makers, they often start feeling a need to measure sides and angles. In fact, after a while, you might even see some students trying to use a ruler or protractor to measure parts of Shape Makers. For instance, they might start believing that all the sides are equal in a Rhombus Maker—and use a ruler to verify their conjecture. This is great! But if they want a better way to measure sides, tell them that one will be introduced in an upcoming Exploration.

Students may also start talking about parallelism and symmetry. Don't mention these concepts in class discussions until students seem to want to use them in communicating about shapes. Then ask students what they mean by these terms. Future Explorations will introduce students to specific computer tools for dealing with the concepts of parallelism and symmetry.

Class Discussion

➡ Have students discuss the questions on the student sheet: Which Shape Maker can make the *least* number of different types of shapes? Why? Which Shape Maker can make the *greatest* number of different types of shapes? Why?

Most students will see that the Square Maker can only make squares. At this point in their exploration, however, they will probably not be familiar enough with the Shape Makers to understand all the different shapes that each Shape Maker can make. That is okay. The objective is to engage students in examining the Shape Makers, and in reflecting on how the Shape Makers operate.

Use of Language

➡ To minimize confusion during discussion, be sure to explain to students the difference between a Shape Maker and the shapes it makes.

See "Teaching Note: Shapes Versus Shape Makers" on page 25. In all communications, it is essential for you and for students to distinguish between a Shape Maker and a shape that it makes.

Activity: Can You Make It?

Use Student Sheets 2 and 3, and sketches:

Mac: Windows:

📁 Quadrilateral Exploration 1   📁 QuadExp1

    ◈ Can You Make It? #1   ◈ CYMI_1.gsp

    ◈ Can You Make It? #2   ◈ CYMI_2.gsp

Students Work in Pairs

➡ Distribute the student sheets Can You Make It? #1 and Can You Make It? #2 to the class.

➡ Explain that students should open the appropriate sketch and use the seven quadrilateral Shape Makers to make the pictures that appear in the middle of the screen (which are also shown on the student sheets).

Students make the shapes in the pictures by manipulating the Shape Makers then placing them on top of the given shapes. There are several different solutions for both of these student sheets.

➡ Using a computer connected to an overhead display or gathering all students around a single computer, demonstrate that the pictures in the middle of the screen may be inadvertently selected or moved.

Using the sketch Can You Make It? #1, click on the picture in the middle of the screen—it will then appear in a black rectangular region, indicating that it has been selected. Also illustrate that this picture can be dragged to another location. Because students may select or move the picture inadvertently while trying to manipulate the Shape Makers, you need to explain what is happening and how to avoid any unwanted effects.

For instance, sometimes, while students are trying to move one of the control points of a Shape Maker that has been placed on the picture, they accidentally select the picture. To deselect it, they need to move the arrow to a location outside the black rectangle and click. If students accidentally move the picture, they should simply move it back to where it was.

➡ When students have completed a student sheet, have them list their answers (letter of shape matched to Shape Maker) on a classroom chart, as shown in the example here. Pairs of students can then attempt to validate or refute other pairs' answers.

	Jon, Alex	Brittany, Laura
	Letter of shape in picture	Letter of shape in picture
Square Maker	D	F
Rectangle Maker	A	A
Parallelogram Maker	C	B
Kite Maker	G	E
Rhombus Maker	F	D
Trapezoid Maker	E	C
Quadrilateral Maker	B	G

➡ As you walk around the room, notice what kinds of things students try and how they are thinking about the Shape Makers and shapes.

See "Teaching Note: Early Reasoning about Shape Makers" on page 26 for a more detailed discussion of the different ways students will be reasoning. Also note what kinds of errors students make and what types of difficulties they have. You might observe students having manipulation and conceptual difficulties.

Manipulation Difficulties

Sometimes students have difficulty placing a Shape Maker on top of a shape it will make. This often happens as students attempt to place a control point of the Shape Maker on top of a vertex of the shape. They then try to place another control point of the Shape Maker on top of another vertex in the shape, and so on. Although this strategy sometimes works, it frequently fails because the control points of the Shape Maker don't fit onto the vertices of the shape in the order in which the student is attempting to make them fit. Alternately, the Shape Maker control points might not move in the way the student is trying to move them.

If you see students having this difficulty—if they can't get a Shape Maker to fit onto a shape that you know it will make—there are a couple of suggestions you can make:

1. *Maybe that control point goes on one of the other vertices of the shape.*
2. *Put the Shape Maker next to the shape you are trying to make instead of on top of it. Now try to make the shape. When you think you are really close,* then *put the Shape Maker on top of the shape and make adjustments to make it fit.*

It is useful to suggest the second strategy to the whole class.

Conceptual Difficulties

The activities in "Can You Make It?" are a little more difficult than the first activity, because students must choose wisely which Shape Maker to use for each shape in the picture. For instance, if they use the Quadrilateral Maker to make a square, they will not have it available for a shape that only the Quadrilateral Maker can make.

One difficulty students might encounter is assuming that a Shape Maker will fit on a shape that it cannot make. Students do not yet understand the properties of the Shape Makers and may think, for instance, that the Rectangle Maker can be used to make shape C on Can You Make It? #1 or shape G on Can You Make It? #2. They must try the Rectangle Maker for themselves to determine that it doesn't work. As you walk around the room, if you see students doing something like this, be sure to ask questions that encourage them to try to figure out why the Rectangle Maker won't make these shapes:

Why won't the Rectangle Maker make shape G?

Student answers to such questions will vary considerably in sophistication. One student might say that the Rectangle Maker won't make shape G because the Rectangle Maker won't tilt correctly. Another student might say "It won't work because the Rectangle Maker has square corners [right angles] and shape G doesn't." Accept various answers. But try to get students whose thinking is erroneous to see their mistakes, not by telling them that their ideas are incorrect but by asking them to explain how they got their answers and by posing variations in tasks that expose these errors *to students*. It is also important to get students to describe and reflect on their findings. See "Teaching Note: Nurturing Progress by Encouraging Reflection" on page 27 for more discussion about what you should be looking for in students' work and how you can encourage reflection by asking questions.

Class Discussion

➡ After students have completed both student sheets, have a discussion about what they found.

Using the classroom charts you made for Student Sheets 2 and 3, have pairs of students tell you which Shape Makers they used for each shape, perhaps writing each pair's answer in a different color. Encourage students to object if they don't believe that a given Shape Maker will make a given shape and to clearly explain their reasoning. Have the class attempt to resolve discrepancies.

Another point that you might discuss with students in this lesson or a future one is the implication of not being able to do something they are trying to do. See "Mathematical Note: I Can't Do It, or Is It Impossible?" on page 29.

TEACHING NOTE

Shapes Versus Shape Makers

It is important for you to use, and encourage students to use, clear and precise language; otherwise, the resulting conceptual confusion will inhibit learning. In particular, it is especially important to use language that distinguishes between shapes and Shape Makers. For instance, one student confused the Shape Makers with the shapes they make, saying "You can make rectangles into squares, but you can't make squares into rectangles." More accurately, this statement would be "You can make the Rectangle Maker into a square, but you can't make the Square Maker into a rectangle." The term *rectangle* should be used to refer to a specific type of shape; the term *Rectangle Maker* should be used to refer to the dynamic computer object that can be used to make rectangular shapes on the screen. An analogy might be useful here. A piece of wire can be used to make various rectangular shapes. The Rectangle Maker, like the wire, is the thing used to make particular rectangles. The Rectangle Maker, like the wire, is not a rectangle; it can simply take on various rectangular shapes.

The use of precise language will help students eventually (after many explorations) develop clear distinctions between three related but distinct concepts. First, there are sets or classes of shapes—for example, the set of all rectangles. Second, there are Shape Makers for sets of shapes—for example, the Rectangle Maker. Third, there are examples of sets of shapes—for example, particular drawings of rectangles. These three concepts are related in important ways. Because the Rectangle Maker can make all examples of rectangles and can make only rectangles, a shape is a rectangle if and only if it can be made by the Rectangle Maker (within screen limitations).

Students who distinguish between and properly relate examples, sets, and Shape Makers develop more powerful reasoning. For example, from their observation that the Rectangle Maker can make any shape the Square Maker makes, such students generally conclude that all squares are rectangles. Because these students see the Shape Makers as representations of classes of shapes, they can reflect on actions taken with the Shape Makers and draw conclusions about properties of and interrelationships between classes of shapes. However, students who think of a Shape Maker as a shape may take the fact that the Rectangle Maker can make a square as meaning that a rectangle is a square.

Thus, in addition to promoting clear communication, forcing students to distinguish between the terms they use for shapes and Shape Makers will regularly focus their attention on the conceptual differences between the two. Whether or not they completely understand this conceptual distinction at first, the regular attention to linguistic distinction will help students eventually come to a clear understanding of these related concepts.

These ideas are discussed further in the Explorations dealing with classification of quadrilaterals. But suffice it to say that encouraging the use of precise language in early Explorations will make it much easier for students to construct proper concepts in subsequent Explorations. Attention should be given to this important issue throughout the Explorations.

TEACHING NOTE
Early Reasoning About Shape Makers

The thinking and strategies that students first employ when dealing with the Shape Maker tasks will vary greatly in sophistication. In this example, three students interpret their manipulation of the Square Maker very differently.

Michael: I think maybe you could have made a rectangle [with the Square Maker].

Jon: No; because when you change [the length of] one side, they all change.

Eric: All the sides are equal.

Michael, Jon, and Eric have abstracted different things from their Shape Maker manipulations. Michael found a visual similarity between squares and rectangles, causing him to conjecture that the Square Maker could make a rectangle. Jon abstracted a physical property: When one side changes length, all sides change. Eric made a mathematical conclusion that the sides of the Square Maker are always equal in length.

Especially during early work with the Shape Makers, much of what students discover will refer to their manipulation experiences. For instance, Manuel noticed that, with the Kite Maker, "If I pull one end out, the other end goes out." After trying to make the rectangle on Can You Make It? #2 with the Square Maker, he said that it couldn't be done: "The Square Maker would only get bigger and twist around—so it can't make a rectangle." Manuel was discovering properties or rules for the Shape Makers, but these rules were not the typical mathematical properties of shapes. Instead, his knowledge was visual-holistic and manipulation-based.

Similarly, Stefanie was considering whether the Parallelogram Maker could be used to make the trapezoid at right.

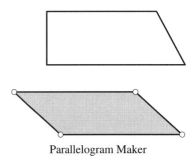

Parallelogram Maker

She did not know enough about the properties of parallelograms to predict that this task was impossible. However, as she manipulated the Parallelogram Maker, she discovered something special about parallelograms that enabled her to solve the problem.

Stefanie: [Pointing to the nonhorizontal pair of opposite sides] No, it won't work. See this one and this one stay the same, you know, together. If you push this one [side] out, this one [the opposite side] goes out. . . . This side moves along with this side.

Like Manuel, Stefanie discovered something special about the characteristic movement of a Shape Maker. This characteristic is a consequence of the constraint on the Parallelogram Maker that its opposite sides stay parallel and congruent. However, Stefanie did not conceptualize her discovery in terms of formal mathematical concepts such as parallelism and congruence. Instead, she was thinking about constraints on possible movements of the Parallelogram Maker. Eventually, she will elaborate on her observations and come to think of this idea in terms of formal geometric concepts.

TEACHING NOTE

Nurturing Progress by Encouraging Reflection

Sustained growth of students' mathematical ideas occurs in a nurturing environment in which students' ideas are always considered worthy of examination but are constantly challenged and reflected upon. Often, the most important role for a teacher is to focus students' attention and reflection on appropriate and potentially fruitful ideas. In the episode below, the teacher is interacting with one student who happens to be working at her computer alone because her partner is absent. She is working on the student sheet Can You Make It? #1.

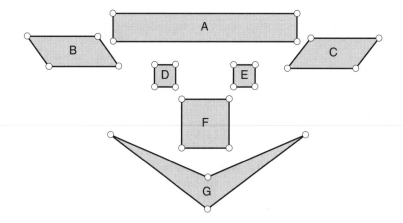

Teacher: Tell me what you're working on.

Naomi: I'm trying the Rhombus Maker on shape B. It's not working because it's leaning to the right and shape B is leaning to the left. Shape C leans to the right, so I'm going to try the Rhombus Maker there. [Putting the Rhombus Maker on shape C and trying to pull it in different ways to make the shape] I don't think that this is going to work.

Teacher: Why are you thinking that?

Naomi: When I try to fit the Rhombus Maker on the shape, and I try to make it bigger or smaller, the whole thing moves. It will never get exactly the right size. [While trying to make shape C with the Rhombus Maker, she makes a shape that resembles a square.] Let's see if I can make a square with this. [Doing so] Here's a square. I guess it could maybe be a square.

Teacher: What could?

Naomi: The Rhombus Maker. But I'm not sure if this is exactly a square. [Pulling the Rhombus Maker off shape C] I don't think it is exactly a square.

Teacher: Why?

Naomi: It's sort of leaning. The lines are a little diagonal. [Continuing in her attempts to make a square with the Rhombus Maker, and to make its sides vertical and horizontal] Yeah, I think this is a square, maybe. [Taking the square-shaped Rhombus Maker and dragging it over to shape E, adjusting it to match] Yeah, it works for E.

Teacher: When you tried to fit the Rhombus Maker on C, did you notice anything about it or about shape C?

Naomi: The Rhombus Maker could make the same shape pretty much, but if you tried to make it small enough to fit on C, it would make the whole thing smaller.

Teacher: You said the Rhombus Maker could make the same shape. What do you mean by that?

Naomi: The Rhombus Maker could make this shape [C], the one with two diagonal sides and two straight sides that were parallel. I don't really know. It could have been almost that shape [C], and it got so close I thought it was that shape. See, it could make the same shape as that [shape C]. [Naomi takes the Rhombus Maker off shape E and manipulates it to make rhombuses with nonright angles, then rhombuses with right angles, then rhombuses with nonright angles. She seems to be trying to get the Rhombus Maker to make an elongated shape.]

Oh, I see why it didn't work! It's because the four sides [of the Rhombus Maker] are even, and this [shape C] is more of a rectangle.

Teacher: How did you just think of that?

Naomi: All you can do is move the Rhombus Maker from side to side and up. But you can't get it to make a rectangle. When you move it this way, it is a square [moving one control point], and you can't move it up to make a rectangle. And when you move this [other control point], it just makes a bigger square.

Teacher: So what made you notice that?

Naomi: Well, I was just thinking about it. If it was the same shape, then there is no reason it couldn't fit onto C. But I saw when I was playing with it that whenever I made it bigger or smaller, it was always like a square, but sometimes it would be leaning up, but the sides are always equal.

Analysis. This episode clearly shows how students' manipulation of the Shape Makers and their *reflection* on that manipulation can enable them to move from thinking holistically about shapes to thinking about shapes in terms of interrelationships between their parts, that is, their properties. By reflecting on why the Rhombus Maker could not make shape C, Naomi increased her understanding of it, developing a property-based conclusion that the Rhombus Maker always has four equal sides. The teacher played an important role in this episode by (a) encouraging Naomi to investigate something that puzzled her and (b) encouraging Naomi to reflect not just on what she was doing, but on what she was thinking.

MATHEMATICAL NOTE

I Can't Do It, or Is It Impossible?

It is common in mathematics to be stumped by a problem—no matter how hard we try, we can't find a solution. But what should we conclude from this? It is vital to keep in mind that our failure may be due to one of two very different causes. On the one hand, it may be that a solution is possible, but our approach to solving the problem is wrong. On the other hand, it may be that there is no solution to the problem. Unfortunately, we often don't know which situation we are in; just because we cannot solve the problem doesn't mean that it is impossible to solve, as many students conclude.

It is important to discuss these two possibilities with students. If they can't show that one of their conjectures is true, they should ask themselves, "Is this task possible?" and look for reasons why it might be impossible. On the other hand, just because they can't show that one of their conjectures is true does not mean that it is false. They may need to search for a different approach to the problem.

Shape Maker Games

Summary

Students explore which shapes can be made by the various quadrilateral Shape Makers, first by playing a game in which one student tries to use a Shape Maker to make a shape made by another student with another Shape Maker, then by manipulating Shape Makers without names to discover which Shape Maker is which.

Mathematical Objectives

Students think more carefully about the kinds of shapes each Shape Maker can make. They start interrelating Shape Makers, seeing, for example, that the Rectangle Maker can make any shape the Square Maker can make. They start taking a more analytic approach to thinking about shapes and Shape Makers, focusing more on the parts and characteristics of the shapes than on the shapes as wholes.

Mac:

📁 Quadrilateral Makers

    📁 Quadrilateral Exploration 2

        📄 Quadrilateral Makers

        📄 Hiding Shape Makers 1

Windows:

📁 QuadMkrs

    📁 QuadExp2

        📄 Quads.gsp

        📄 Hide_SM1.gsp

Required Materials

Session	Student sheet	SS#	Geometer's Sketchpad sketch
1	Shape Maker Challenge Game	4–6	Quadrilateral Makers (Mac) Quads.gsp (Windows)
2	Identify the Hiding Shape Makers #1	7	Hiding Shape Makers 1 (Mac) Hide_SM1.gsp (Windows)

For session 1, you will need a Shape Maker Selector for each pair of students. There are several ways you can create this device:

1. To the six faces of a wooden inch cube, paste or tape the following names: Square Maker, Rectangle Maker, Parallelogram Maker, Kite Maker, Rhombus Maker, Trapezoid Maker. You can use a computer to print (or you can have students neatly write) these names on computer labels; then cut out the names and attach them to the cubes.
2. Cut out the cube pattern on the Shape Maker Selectors sheet (page 221); then fold and tape it to make a cube.
3. Make a spinner from the spinner pattern on the Shape Maker Selectors sheet and a bent paper clip. Students put their pencil tip at the center of the circle, inside the paper-clip loop, then spin the paper clip to select a Shape Maker.

SESSION 1

Activity: Shape Maker Challenge Game

Use Student Sheet 4–6, and sketches:

Mac:

Quadrilateral Exploration 2

Quadrilateral Makers

Windows:

QuadExp2

Quads.gsp

Class Discussion

➡ Explain the Shape Maker Challenge Game to students.

The game is for two players. The player who goes first in a game is called Player 1. (Students take turns going first in games.) To start a game, Player 1 uses the Shape Maker Selector to randomly choose the Shape Maker he or she will use for this game, then writes his or her initials next to that Shape Maker in the Player 1 column for Game 1 on the student sheet. Player 2 then uses the Selector to randomly choose a Shape Maker different from that of Player 1. Player 2 puts his or her initials next to the selected Shape Maker in the Player 2 column for Game 1.

To play, Player 1 makes a shape with his or her Shape Maker. Player 2 then tries to make the same shape with his or her Shape Maker. If Player 2 successfully makes Player 1's shape, then Player 2 tries to make a shape that Player 1 can't make. The players keep taking turns until one player cannot make the shape made by the other player. The player who makes a shape that his or her opponent cannot make is the winner. Students circle the initials of the winner, then draw the shape that could not be made and explain why it couldn't be made.

➡ Explain the *size rule* to students.

To ensure that the shapes students make can be clearly seen, students must follow the size rule: (Part 1) The shapes that students make with their Shape Makers must be small enough to fit completely on the screen. (Part 2) The shapes must be large enough so that the circle formed by the flat end of a standard pencil eraser will fit inside.

➡ Illustrate how to play the game.

It's best to illustrate by playing the game in front of the class. You can be one player, a student the other.

Illustration 1. The teacher (T) is Player 1. She uses the Shape Maker Selector and gets the Rectangle Maker. Player 2 (the student, KB) gets the Rhombus Maker. Player 1 begins and makes the shape at right with the Rectangle Maker:

Rectangle Maker

Player 2 must try to make the same shape with the Rhombus Maker (which can't be done). Thus, Player 1 wins. (Note that if Player 1 had made a square, Player 2 could have made it with the Rhombus Maker. Player 2 could then have made a shape that Player 1 could not make—one without right angles—and so Player 2 could have won the game.)

Illustration 2. Player 1 (now the student) uses the Shape Maker Selector and gets the Square Maker. Player 2 (the teacher) gets the Parallelogram Maker. Player 1 begins and makes the shape at right with the Square Maker:

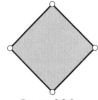

Square Maker

Player 2 must now make the same shape with the Parallelogram Maker, which she does. Player 2 now makes a new shape with the Parallelogram Maker:

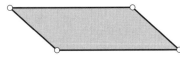

Parallelogram Maker

Player 1 must try to make this shape with the Square Maker. Because Player 1 cannot make this shape with the Square Maker, Player 2 wins.

The above games would be recorded on the student sheet as shown:

Game	Player 1	Player 2	Draw the shape that couldn't be made	Why couldn't this shape be made?
1	Square Maker Rectangle Maker ⓣ Parallelogram Maker Rhombus Maker Trapezoid Maker Kite Maker	Square Maker Rectangle Maker Parallelogram Maker Rhombus Maker ⓚⒷ Trapezoid Maker Kite Maker	▭	The Rhombus Maker couldn't make different side lengths.
2	Square Maker ⓚⒷ Rectangle Maker Parallelogram Maker Rhombus Maker Trapezoid Maker Kite Maker	Square Maker Rectangle Maker Parallelogram Maker ⓣ Rhombus Maker Trapezoid Maker Kite Maker	▱	The Square Maker only makes squares. This shape isn't a square.

Students Work in Pairs

➡ Distribute the Shape Maker Challenge Game student sheets and the Shape Maker Selectors to the class.

Class Discussion

➡ After each pair of students has played five or six games, have a discussion about what happened.

Pairs of students should describe the results of one of their games, clearly explaining why one Shape Maker could not make the shape made by another Shape Maker.

➡ As students explain their games, encourage them to think more deeply about the reasons why one Shape Maker won't make another.

For instance, one student said that the Rectangle Maker wouldn't make the chevron shown at right because the Rectangle Maker only made rectangles and squares. You can encourage students to move beyond such a holistic description to thinking about properties of shapes by asking "What is it about the Rectangle Maker that allows it to make rectangles and squares but not this shape?"

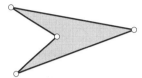

Keep in mind that, as students start focusing on particular visual characteristics of shapes, they may describe these characteristics quite informally. For example, one student replied to the question suggested above by saying "The Rectangle Maker's sides are always straight. In this shape [the chevron], the sides are not straight." In this case, it is worthwhile to ask the student what *straight* means. But do not force students to use formal mathematical concepts such as right angles and perpendicularity at this point—these concepts will come later in the unit. What's important now is to have students start thinking about the components of the shapes (such as sides and angles) and how they are related (straight, perpendicular, not slanted, at right angles).

Help students' thought become more sophisticated by asking questions that encourage them to reflect on their own thinking, especially where that thinking may be erroneous. See "Teaching Note: Perturbation and Progress" on page 35 for an illustration.

Use of Language

As students describe their ideas, do not impose formal mathematical language on them. However, to make communication possible, students need to think about and agree on exactly what they mean by some of the terms they are using. For instance, students often use the term *straight* to mean perpendicular, vertical, horizontal, or "not curvy." Because these meanings are so different, some classroom consensus needs to be reached about the use of the term. See "Teaching Note: Establishing Classroom Consensus on Language" on page 36 for an illustration of how such a discussion might proceed. You might even post a language reference chart in the classroom, listing terms that students are using and the meanings they agree to give those terms.

➡ Ask students whether who goes first in a game matters.

If students do not see that order matters, give as an example the situation in which one player selects the Kite Maker and the other the Rectangle Maker. Have students show examples of how the player with the Kite Maker or with the Rectangle Maker can win, depending on who goes first.

➡ Ask students what they learned about various Shape Makers by playing this game. Ask them what questions they have about the various Shape Makers.

SESSION 2

Activity: Identify the Hiding Shape Makers 1

Use Student Sheet 7, and sketches:

Mac:

📁 Quadrilateral Exploration 2

🔲 Hiding Shape Makers 1

Windows:

📁 QuadExp2

🔲 Hide_SM1.gsp

Students Work in Pairs

➡ Distribute the Identify the Hiding Shape Makers 1 student sheet to the class, and explain the task.

Students manipulate the Shape Makers without names in the sketch Hiding Shape Makers 1 to discover which Shape Maker is which. For each hiding Shape Maker, they explain how they determined its identity.

Class Discussion

➡ After students have decided which Shape Maker is which, have a discussion about what they found.

Students should describe not only the reasons they wrote for their identifications, but also some of their thoughts and any mistakes they made as they were trying to identify the Shape Makers.

➡ This is a good time to have students discuss their Conjectures and Queries sheets. What have they found? How did they find it?

Assessing Student Progress

➡ Listening to students' thoughts about the problems they encounter in this activity, as well as reading what they write on their student sheets, can help you assess how students' knowledge about Shape Makers is developing.

See "Teaching Note: Control Points, Shapes, and Properties" on page 37 for a discussion of typical levels of sophistication in students' reasoning about this task.

Homework

 Have individual students write answers to these questions: "In the Shape Maker game, suppose I roll the Parallelogram Maker and you roll the Kite Maker. If you go first in the game, who should win? Why? If I go first in the game, who should win? Why?"

By reading students' responses, you can gain a good sense of how they are thinking about these ideas. Keep in mind, though, that many students will not yet be familiar enough with the Shape Makers to fully understand the possibilities.

Depending on students' responses, you might want to give them time to check their answers on the computer the following day. After doing this, you might also want to have the class discuss their answers.

TEACHING NOTE

Perturbation and Progress

Michael and Jon have concluded that all shapes made by the Rhombus Maker have four sides and four corners, have two sets of parallel lines, and are closed figures. The boys, however, have also decided that the Rhombus Maker and the Parallelogram Maker are the same, that they have the same properties. On hearing the boys' conversation, the boy's teacher asks questions that encourage them to reflect further on their idea.

Teacher: Can every shape made with the Rhombus Maker be made by the Parallelogram Maker?

Boys: Yes.

Teacher: Can every shape made with the Parallelogram Maker be made by the Rhombus Maker?

Boys: Yes.

The teacher then makes several shapes with the Rhombus Maker, and the boys make those shapes with the Parallelogram Maker. He then makes a non-equilateral parallelogram with the Parallelogram Maker and asks the boys if they can make it with the Rhombus Maker. Michael tries for several minutes.

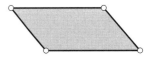

Parallelogram Maker

Michael: Oh, I see. The Rhombus Maker has all sides equal, like a square. A parallelogram is like a rectangular diamond. A rhombus is like a square diamond.

In this case, the teacher saw that the boys had made an incorrect conclusion. He did not want to simply tell the boys that they were wrong—he wanted them to discover this themselves. So the teacher posed problems that, through the boys' own exploration, could help them see their error. The fact that Michael could not make the nonequilateral parallelogram with the Rhombus Maker caused a perturbation in his theory about the relationship between the Rhombus Maker and the Parallelogram Maker. Michael resolved this perturbation by analyzing the Rhombus Maker and subsequently discovering that it, unlike the Parallelogram Maker, has to have all sides congruent. Consequently, Michael developed a more sophisticated understanding of these two Shape Makers.

This discovery was an important step in the boys' progression to more sophisticated geometric thought. In fact, students at the visual level may say—and indeed only see—that the Rhombus Maker and the Parallelogram Maker "make the same shape." They simply see two figures with the same overall shape. Michael's analysis moved him a step closer to thinking about these shapes in terms of their properties.

Establishing Classroom Consensus on Language

As students talk about the Shape Makers, you will notice that they use terms in inconsistent ways. One common example is their use of the term *straight*. In the next episode, a teacher gets students first to clarify the different meanings they have for *straight,* then to agree on a common meaning. Such agreement is a prerequisite for clear communication among students.

Teacher: What do you think we should do about that word *straight?*

Carolyn: Well, um, I think that it basically means like [puts her hands in a vertical position] just any straight line.

Teacher: Naomi, what do you think we should do with the word *straight?*

Naomi: I think we have some people who think that we should use it when, like, it's up and down [puts a pencil in a vertical position] or straight across [puts a pencil in a horizontal position]. And some people think that *straight* is like straight [puts a pencil in a vertical position] or not wiggly or something. So, I think for right now we should just try not to even use *straight*.

Brenda: I think that we should use *straight,* but we could use it as long as the line is not wavy; it can be diagonal [puts her hand in an oblique position]. But maybe when somebody means straight and not diagonal, they'll say the lines are straight, not diagonal.

Teacher: So, you're thinking that when a line is horizontal [puts a ruler in a horizontal position], vertical [puts another ruler in a vertical position], or diagonal [puts a ruler in an oblique position], we can call them straight. And then we know that any straight line is one that is not curved or wavy.

Brenda: Yeah!

Linnea: We should just do what Carolyn said and say, like, straight. But then after you say straight you have to say, like, show what way it is pointing. It's not straight if it's like this [moves her hand obliquely], or this is not straight [makes a wavy motion].

Teacher: Okay. So every time somebody uses the word *straight* you think we should have a definition to go with it, like a hand motion.

Jarod: I think that we should use the word *straight* as up and down [uses his arm and puts it in a vertical position] or like this [uses his arm and puts it in a horizontal position], not a diagonal [puts his arm obliquely]. Because with the Shape Makers, none of the lines have been wavy.

Carolyn: I sorta agree with Linnea, because then you can use *straight* a lot more than if we would've just said not wavy.

Teacher: If I drew a line coming across [runs her finger across the top of the ruler she is holding in a horizontal position] using this as my guide, this is straight, right? If I took that and changed the orientation of it [putting the ruler in an oblique position], do I still have a straight line? [Some students nod yes.]

Juan: Yeah!

Li Chen: Well, I agree with Linnea a lot, because people might believe different things about *straight* and so they'll know which way they believe in after they use their hand motion.

Teacher: So do I hear as a class, right now, you want the word *straight* to always be followed by a hand motion? [Li Chen is nodding yes.] Or do I hear that you want the word *straight* to mean any line that isn't curved or rippled, regardless of how it is oriented? [Several students nod yes.]

Class: Yes!

Teacher: We have to have an agreement so that when we're talking with each other we know what we're talking about. So that you are comfortable at this point, we'll go with whatever the majority of the students are thinking.

How many of you want the word *straight* to be not curved or wavy, regardless of the orientation? [Most hands go up.]

How many want to always have a hand motion with it? [Four students raise their hands.]

Now here's an important question. Hand motion people, for our discussions can you picture a straight line being any line that isn't curved, no matter what its orientation? [These four students say yes.]

Okay. Then as a class, we'll agree that *straight* means not curved or wavy, no matter what the orientation of the line is.

TEACHING NOTE

Control Points, Shapes, and Properties

The reasoning you will observe students using in identifying the Hiding Shape Makers will vary in sophistication both between pairs and as single pairs move from one Shape Maker to another. For instance, one pair of students, Martin and Ted, manipulated Shape Maker A and said immediately that it was the Rhombus or Parallelogram Maker "because of the way it moves." Martin then decided that it was the Rhombus Maker, because he mistakenly thought that they had previously discovered that the Parallelogram Maker couldn't make rectangles or squares. Ted countered that it couldn't be the Rhombus Maker, because they had found that the Rhombus Maker had to have all of its sides equal.

Martin and Ted next concluded that Shape Maker B was the Rectangle Maker, because it could only make rectangles and squares. After manipulating Shape Maker C into a chevron, they said it was the Kite Maker, because it was the only Shape Maker that could make that shape. They also concluded that Shape Maker E was the Rhombus Maker, because it had four "even" sides.

Other students working on this task mentioned the Shape Makers' control points: "[Manipulating one of the control points of G] It can't be the Quadrilateral Maker, because the Quadrilateral Maker doesn't have a control point that makes the whole shape move." Another student recalled specific configurations that certain Shape Makers had made, saying of E, "It can't be the Trapezoid Maker because it can't make the original trapezoid shape [meaning the shape that the Trapezoid Maker starts as when the Shape Maker sketch is first opened]."

Thus, there seem to be three levels of sophistication in students' explanations. At the lowest level are explanations that deal with the movement of a Shape Maker or its control points. At the next higher level, students say that the given Shape Maker can only make certain shapes—the Rhombus Maker can't make (nonsquare) rectangles. At the highest level are explanations that refer to typical geometric properties— all the sides are equal (i.e., congruent) in the Rhombus Maker. Of course, each of these explanations may be correct. But, according to the van Hiele hierarchy, the third and final level of response is more sophisticated because it expresses a relationship between the parts of a shape. Note, however, that because the sides and angles of the Shape Makers are not measured yet, judgments about equality can be only visual in nature.

You can encourage students to make their thinking more sophisticated by asking questions such as "What is it about the Rhombus Maker that prevents it from making rectangles?" or "Why do you think the Kite Maker moves the way it does?"

Predict and Check

Summary

Students investigate the types of shapes that can be made by the measured Quadrilateral Makers, which display the measures of angles and side lengths.

Mathematical Objectives

Students use the measured Shape Makers and the concepts of angle and length measurement to more precisely describe the operation of the quadrilateral Shape Makers. They start to formulate standard mathematical properties of geometric shapes, that is, characteristics that express relationships between parts of shapes.

Mac:

📁 Quadrilateral Makers

   📁 Quadrilateral Exploration 3

      📄 Hiding Shape Makers 2

      📄 P&C Quadrilateral Maker

      📄 P&C Kite Maker

      📄 P&C Parallelogram Maker

      📄 P&C Rectangle Maker

      📄 P&C Rhombus Maker

      📄 P&C Square Maker

      📄 P&C Trapezoid Maker

Windows:

📁 QuadMkrs

   📁 QuadExp3

      📄 Hide_SM2.gsp

      📄 PC_Quad.gsp

      📄 PC_Kite.gsp

      📄 PC_Para.gsp

      📄 PC_Rect.gsp

      📄 PC_Rhomb.gsp

      📄 PC_Squa.gsp

      📄 PC_Trap.gsp

Required Materials

Session	Student Sheet	SS#	Geometer's Sketchpad sketch
1–3	Predict and Check	8–10	All P&C Shape Makers
4	Identify the Hiding Shape Makers 2	11	Hiding Shape Makers 2 (Mac) Hide_SM2.gsp (Windows)

Introducing Shape Maker Measurements

Class Discussion

➡ Using a computer connected to an overhead display or gathering all students around a single computer, explain the Shape Maker measurements in the (Predict and Check) sketches.

There is a separate sketch for each quadrilateral Shape Maker. Open the sketch P&C Quadrilateral Maker. Point out that there are measurements shown on the screen, as well as letters labeling the four vertices of the Quadrilateral Maker (see Figure QE 3.1). The measurements indicate the lengths of the sides and the measures of the angles in this Shape Maker. Illustrate that, as you drag the control points of the Quadrilateral Maker from one location to another, some of the measures change—indicating the measurements of the new quadrilateral you have made.

Length(AB) = 60 pixels
Length(BC) = 55 pixels
Length(CD) = 158 pixels
Length(AD) = 73 pixels

Angle(A) = 123°
Angle(B) = 161°
Angle(C) = 35°
Angle(D) = 41°

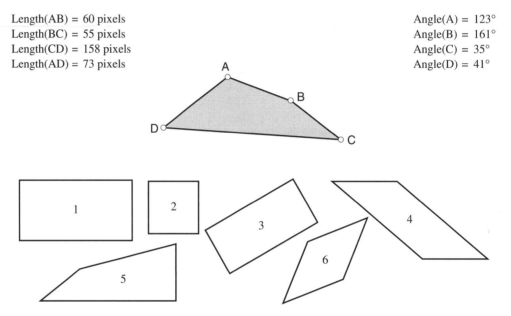

Figure QE 3.1. P&C Quadrilateral Maker screen

➡ Ask students what they think the various measurements mean. Be sure that they can match the measurements to the appropriate sides and angles.

Pointing to the statement Angle(A) = 123° on the screen, ask students:

What does this statement mean?
What does the A mean? Show me where angle A is.
What are degrees?
What does it mean to say that an angle measures 123 degrees?

Pointing to the statement Length(AB) = 60 pixels, ask students:

> *What does this statement mean?*
>
> *Why are two letters used here? Show me where side AB is.*
>
> *Does anybody know what a pixel is?*

See "Mathematical Note: Measuring Sides and Angles" on page 43 and "Mathematical Note: The Meaning of *Degree*" on page 45 for a discussion of these concepts.

Activity: Predict and Check

Use Student Sheets 8–10, and sketches:

Mac:

📁 Quadrilateral Exploration 3

    ◈ P&C Square Maker, etc.

Windows:

📁 QuadExp3

    ◈ PC_Squa.gsp, etc.

Students Work in Pairs

➡ Distribute the Predict and Check student sheets to the class.

In this activity, students predict, then check, with the P&C Shape Maker sketches, which of the shapes drawn on the student sheet can be made by each quadrilateral Shape Maker. If a figure cannot be made by a Shape Maker, they explain why not.

➡ As you walk around the room and interact with students, have them explain what they are doing. Ask questions that focus students' attention on why a figure can or cannot be made by a certain Shape Maker.

For instance, after students make predictions for a Shape Maker, but before they check it, ask them why they made the prediction they did. Students often say, for example, that shape 4 can be made with the Rectangle Maker. Ask why. Some students say that it is just a tilted rectangle, like shape 3. So they conclude that, because shape 3 can be made from the Rectangle Maker, so can shape 4. Other students will be starting to think in terms of the properties of shapes. For instance, they might say that the Rectangle Maker always makes shapes with "square corners" (right angles), so, because shape 4 doesn't have such corners, the Rectangle Maker can't make it.

After students have checked their prediction with a Shape Maker, ask them why they think a certain Shape Maker could or could not make a given shape.

The Value of Predictions. By requiring students to predict, then check their answers, we encourage them to reflect on their conceptions of the Shape Makers. This provides opportunities for students to recognize errors in their conceptions, leading them to revise their conceptions and make them more sophisticated.

Conjectures and Queries. Remind students to continue adding ideas to the Conjectures and Queries sheets, both their own and the class chart.

Class Discussion

➡️ Have students explain which Shape Makers can make which shapes and why.

There are bound to be disagreements about which Shape Makers can make which shapes. Have students talk about their disagreements and try to resolve them. Also solicit several student explanations for why a Shape Maker can or cannot make a given shape.

You will see a wide range of sophistication in students' reasoning. At the highest level, students will give verbal, logical arguments dealing with properties of shapes: "The Rectangle Maker can't make shape 4 because rectangles always have four right angles, and shape 4 doesn't have any right angles." At the next lower level, students will say, for example, that the Rhombus Maker can make shape 6 because the figure is a rhombus. (If so, ask students how they know it is a rhombus.) At the lowest level, students will say simply that they were able to make shape 6 with the Rhombus Maker. (If so, ask students why they were able to make this shape with the Rhombus Maker.)

Often, students will have a difficult time saying exactly why a shape can or cannot be made by a particular Shape Maker. Ask such students questions that prompt them to clarify their meanings and go beyond vague statements that deal with shapes holistically to statements that focus on properties of shapes. See "Teaching Note: The Rectangle Maker Can't Make a Slant" on page 45 and "Teaching Note: Cultivating Clarity" on page 49 for examples of how you might deal with this difficulty.

Activity: Identify the Hiding Shape Makers 2

Use Student Sheet 11, and sketches:

Mac: Windows:

📁 Quadrilateral Exploration 3   📁 QuadExp3

  🔺 Hiding Shape Makers 2   🔺 Hide_SM2.gsp

Students Work in Pairs

➡️ Distribute the student sheet Identify the Hiding Shape Makers 2 to the class.

Students manipulate the Shape Makers without names in the sketch Hiding Shape Makers 2 to discover which Shape Maker is which. Comparing students' work on this task to their work the first time they did the task will illustrate students' progress in thinking about the Shape Makers.

Measuring Sides and Angles

Length

The length of a line segment is measured by determining how many unit lengths it takes to completely cover the segment with no gaps or overlaps. For instance, to say that a segment is 5 centimeters long means that it takes 5 unit lengths of 1 centimeter to cover the segment.

There are many different kinds of unit lengths. The most familiar are feet, inches, miles, meters, kilometers, centimeters, and millimeters. However, on the computer—since we want to work as much as possible with whole numbers—we use a very small unit length called a *pixel*. The 5-cm segment above measures 157 pixels (1 millimeter is approximately 3.14 pixels.)

Angles

Angles are measured in degrees. Think of one degree as the small amount of rotation it takes to move one ray onto the other as shown below.

It takes 360 one-degree rotations to make a complete revolution, all the way around a circle. Several degree measures are shown below. For example, it takes 5 one-degree rotations to make a 5° angle, 10 one-degree rotations to make a 10° angle, and 30 one-degree rotations to make a 30° angle.

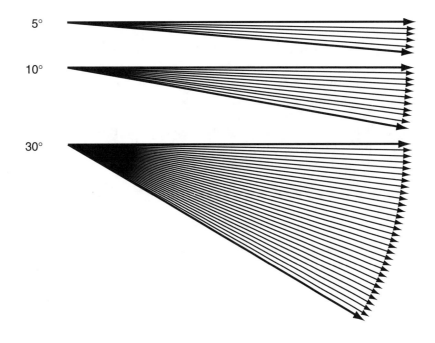

The degree measure of an angle tells us how much to rotate one side of the angle, so that it overlaps the other. This makes measuring most angles simple, as we can see below:

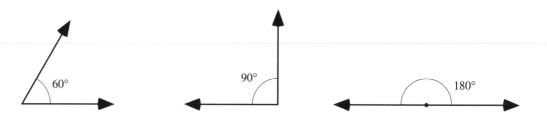

But one troublesome point arises in angle measure. When two rays intersect at their endpoints, there are two possible rotations that move one onto the other, so there are two possible angle measures. For instance, both measurements of the following angle make sense:

The fact that are there two sensible ways to measure an angle can cause confusion. If we ask what the measure of the above angle is, which measure should we give? In many contexts, mathematicians resolve this difficulty by saying that all angles must have measure less than or equal to 180°. This is the convention used by The Geometer's Sketchpad.

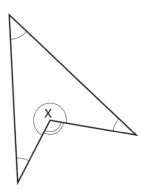

However, this convention sometimes causes difficulty when we analyze angles of polygons. For instance, the well-known theorem "The sum of the angles in a quadrilateral is 360°" refers to *interior* angles of quadrilaterals. The interior angles are inside or, in the interior of, a quadrilateral—as indicated by single arcs in the diagram at right. Because the interior angle at X—the one with a single arc—is greater than 180°, Sketchpad gives the measure of the angle indicated by the double arc.

Thus, if we use the measured Quadrilateral Maker to make this shape, we will not find the total measure of the angles to be 360°. To find the measure of the interior angle at X, we must subtract the displayed Sketchpad measure from 360. When this value is added to the measures of the other interior angles, a sum of 360 will result.

The Meaning of *Degree*

The division of a circle into 360 parts is ancient, a degree being the amount of rotation of the sun about the Earth in a day according to ancient Babylonian and Egyptian computations.

Some students wonder why the term *degree* is applied both to angles and to temperature. Examining the derivation of the word in a dictionary can clear up the confusion. According to the dictionary, the word *degree* means:

1. a step in an ascent or descent;

2. a step or stage in a process, especially, one in an ascending or descending scale;

3. a step or stage in intensity or amount—the relative intensity, extent, measure, or amount of a quality, attribute, or action;

4. in geometry, a unit of angle or arc derived by dividing a complete revolution (circle) into 360 segments; an angle equal to the 90th part of a right angle;

5. in thermometry (measure of temperature), a unit of difference in temperature on a temperature scale (the interval between the freezing and boiling points of water being divided into 180 equal parts in the Fahrenheit scale and 100 parts in the Centigrade or Celsius scale);

6. in algebra, the rank of an equation or expression as determined by the highest power of the unknown or variable.

These meanings are directly derived from the roots of the word: in Middle English, *gree, gre*—step in a series; in Latin, degree = *de* (down) + *gradus* (step)—step down. Related words are *grade*—step or stage in a process; *gradual*—having steps; *gradient*—rate at which something rises.

The Rectangle Maker Can't Make a Slant

Two students, Manuel and Toshi, are trying to make the shapes on student sheet Predict and Check with the measured Rectangle Maker.

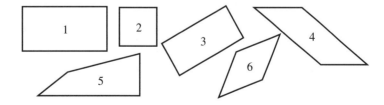

Manuel: The answer for shape 4 is no, because the Rectangle Maker can't make a slant.

Toshi: Yes it can. Shape 3 is in a slant, and we made it with the Rectangle Maker.

Manuel: The Rectangle Maker has to have straight lines.

Toshi: They are straight lines. All of these are straight lines.

Manuel: Yeah, they are straight, but on a slant.

Toshi: [Toshi makes a rectangle with the Rectangle Maker.] This is at a slant right now.

Manuel: The Rectangle Maker can't make something that has a slant at the top and stuff, like shape 6. I know the lines are straight, but they are at a slant, and it [the Rectangle Maker] always has lines that aren't at a slant.

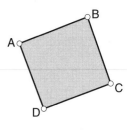

Toshi: Do you mean it has to have a straight line right here? Like coming across? [See the figure at right.]

Teacher: [Overhearing the boys' conversation] What do you mean at a slant, Manuel?

Manuel: It's going in a diagonal direction. See number 4. [He brings down the Rectangle Maker next to shape 4.]

Toshi: Are you talking about the whole Shape Maker, or are you talking about just this line [side AB]?

Manuel: Just that line.

Toshi: I thought he was talking about the whole Shape Maker.

Manuel: No, they are straight lines, but they are at a slant.

Teacher: Are the lines straight or slanted in shape 5?

Manuel: See how shape 5 is at a slant there [upper left vertex]. The Rectangle Maker can't make something that is pointing down or up.

Teacher: Are there parts of this shape that the Rectangle Maker could make?

Toshi: Yeah! It can make a right angle.

Teacher: So what do you wish we could change in this shape so the Rectangle Maker could make it?

Manuel: [Pointing to the left side of shape 5] I wish this side here would go over to here. These two [the left and right sides] should be parallel. We want the top to go straight across.

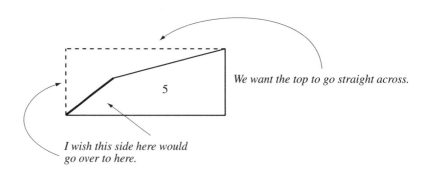

We want the top to go straight across.

I wish this side here would go over to here.

Teacher: Why isn't the top coming straight across in Shape 5?

Toshi: The control point. You have to move this control point [upper left] up here to get it straight.

Teacher: If we moved that control point, what would we create?

Manuel: Something that looks a little bit longer, like shape 1.

Toshi: A right angle.

Teacher: You said that it created a right angle.

Toshi: Yeah, because there would be a straight line right here [the left side] and a straight line right here [the top].

Manuel: I agree with Toshi.

Toshi: And that would be a right angle. It would be exactly like this [pointing to the right angle at the bottom right of shape 5].

Manuel: It would be like that but upside down.

Teacher: Okay. You get a right angle, and what effect does that have on the lines?

Manuel: It makes them all be, like, none of them will be at a slant.

Toshi: So are you saying if there is one right angle?

Manuel: None of the two sides will be at a slant [drawing an acute angle]. There could be other ones [sides] in the shape that could be, if it was just like a regular corner of a shape, like this [circles the bottom left angle of shape 3], then it would be a right angle.

Teacher: So do you think, Manuel, every time you have a right angle, you have lines that are not slanted?

Manuel: Yes.

Toshi: Yeah, you do. Well, slanted. What do you mean slanted? Do you mean like at a diagonal?

Manuel: Slanted means like this [draws an oblique segment].

Teacher: So what are you really talking about, Manuel?

Manuel: The shape being slanted.

Teacher: Maybe it would help us understand if we can figure out exactly what is different about shapes 3 and 4.

Toshi: If this line would be pushed up [motioning to the right side of shape 4], then it [the new shape] would be the same shape as shape 3.

Manuel: These two lines on shape 4 [pointing to the horizontal top and bottom] are not exactly on top of each other.

Teacher: Okay. Keep on really looking at what makes these shapes different. Maybe some of the measurement information on the screen will help you. Watch the numbers up there and see if they help you answer some of the questions. I will return.

[The teacher leaves and Toshi manipulates the measured Rectangle Maker in silence for a minute.]

Toshi: Oh, I just noticed the Rectangle Maker has to always have four 90° angles. And that [shape 4] does not have 90° angles. And so this one [taking the Rectangle Maker down to shape 3] has to have a 90° angle, because we made it with the Rectangle Maker. A 90° angle is a right angle, and this [shape 4] does not have any right angles. So, that's something different that they have. The Rectangle Maker can't make shape 4 because it always has to have right angles.

Manuel: Yeah!

Toshi: And shape 5 doesn't have 90° angles, so therefore you can't make it. Look at the angles.

Manuel: Oh, now I get what her [the teacher's] question was—the angles.

Toshi: What do you mean?

Manuel: Now I get that I am supposed to say angles. The angles control the shape. See, the angles are always 90° [pointing to parts of the Rectangle Maker]. If [angle] A was at a 60° angle, and [angle] B was at a 100° angle, then it would cause it to look something like that [shape 4].

Toshi: Oh. Okay.

Manuel: The first no [for the Rectangle Maker] we get is for shape 4; what are we going to write on our sheet for this? Because of the 90° angles story, or because it is not a rectangle or a square? I like the 90° angles.

Toshi: Yeah, 90° angles. And I'm putting no for shapes 5 and 6 because they don't have any 90° angles, they don't have right angles.

Manuel: Shape 5 does have a right angle, though.

Toshi: Yes. But this [pointing to the Rectangle Maker] always has to have four. So it can't make shape 5.

Manuel: Right.

Analysis. Manuel and Toshi are trying to find a way to describe the fact that the Rectangle Maker can't make shapes that do not have perpendicular sides. However, they have not constructed the concept of perpendicularity yet. So they use terminology that they have available, such as "slanted" and "straight." Unfortunately, these terms don't adequately describe the concept they have in mind. Furthermore, Manuel and Toshi use the terms differently, making communication difficult. But they are committed to trying to communicate with each other.

The teacher does several things to try to help the boys. First, she constantly asks for clarification—"What exactly do you mean by this?" Second, she tries to get the boys to use the concept of angles—in particular, right angles—to analyze the differences in shapes. But even though the boys see that the left side and top of shape 5 need to form a right angle in order for the Rectangle Maker to make it, they are unable to use the concept of right angle to distinguish between shapes 3 and 4. However, though the boys do not yet see that shape 3 has all right angles and shape 4 does not, the teacher has shifted the boys' focus of attention from viewing shapes as wholes to examining parts of shapes. The boys start to analyze relationships between parts of shapes more carefully. Finally, recognizing that the boys need more time to manipulate and reflect on the Shape Makers, she leaves them. But before doing so, she suggests that the boys look at the measurements of the Shape Makers. She feels that this will help them see the role that right angles play in determining whether the Rectangle Maker can make a shape.

After the teacher leaves, Toshi manipulates the Rectangle Maker, focusing his attention on its measurements. Through this manipulation, he discovers that the Rectangle Maker always has four right (90°) angles. Furthermore, because he has abstracted this property, he is able to use it to analyze the shapes on the student sheet. He sees that the Rectangle Maker can't make shapes 4, 5, and 6 because they don't have four right angles.

The boys have not been forced by the teacher to use the right-angle concept and language. Instead, the teacher has made them see that their current intuitive ideas, though correct, are not adequately expressed by the terms *slantiness* and *straight* and, in fact, can be expressed very nicely using the concept of right angles. The boys adopt the right-angle language because they see its value.

TEACHING NOTE

Cultivating Clarity

At the end of a class discussion of students' work on the student sheets Predict and Check, the teacher has asked what students have discovered or are thinking about the various Shape Makers. Tyrone is describing what he and his partner found out about the Rhombus Maker.

Teacher: Does anybody have an idea that we should write down?

Tyrone: Well, we found out that the rhombus has to have even sides, but the angles can be different. Like in the square, they always have to have four 90° angles. But in the rhombus the angles can be different. They don't have to be, like, specific. They don't have to all be right angles. Because the square always has to have four [pause] 90° angles.

Teacher: You said rhombus and square. Do you mean the rhombus and square shapes?

Tyrone: No, sorry, I mean the Rhombus Maker and Square Maker.

Teacher: So you are saying that the Rhombus Maker has to have equal sides, but its angles don't have to be 90°?

Tyrone: Or the angles don't have to be the same. They can be different.

Teacher: They don't all have to be equal?

Tyrone: No, the angles don't all have to be the same, like in the square.

Teacher: How are "equal" and "the same" different?

Tyrone: Well, they're not, there's not really, well.

Teacher: I'm just trying to understand what you mean. Can you tell me the difference?

Tyrone: Equal is like, the same. Well, you could say "equal" because it's like the same amount. Like, say you have three cookies and someone else has three cookies, you could say you have an equal amount.

Teacher: So "equal" and "the same" are the same?

Tyrone:	Yeah! They're basically the same.
Teacher:	So, Tyrone and Manuel [Tyrone's partner] are conjecturing that the Rhombus Maker has to have the same side lengths, but the angles don't have to be 90°? That is, they don't have to all be the same amount of degrees [writing this statement on the class Conjectures and Queries chart].
Brenda:	Are you saying that all the angles can be different? Can they all be different amounts of degrees?
Tyrone:	Well [looks at the information on his student sheet], let me see. Well, no! I'm just saying that two of the angles can be one kind and the other two can be another kind; like two of them can be 135 and the other two could be 35 or 45.
Teacher:	Do you think that's important? Or can I just say they don't have to be the same? Is it important to clarify that?
Tyrone:	Yeah! Because they can't all be.
Teacher:	What do you wish your statement would say?
Tyrone:	Well, that two angles can be the same, and the other two angles can be the same.
Brenda:	Do you mean two can be the same, or have to be?
Tyrone:	Can be.
Teacher:	Any other conjectures or ideas I should put on our chart?

Analysis. As the students explain their ideas and conjectures, the teacher acts as a recorder and, at times, a clarifier. She encourages the students to elaborate on their meanings, so that those meanings become completely clear. In a classroom culture of inquiry, students understand their obligation to make sense out of what other students have said. They question statements that they don't understand or that they disagree with. In the episode described here, both the teacher and Brenda asked Tyrone for clarification. Tyrone did not feel that he was being criticized or that he had done something wrong—he understood that the clarifications being sought were useful, and that this type of questioning was a natural and important part of the activity in his mathematics class.

Shape Maker Properties

Summary

Students investigate the types of shapes that can be made by the M (measured) Quadrilateral Makers, which display the angle measures and side lengths and provide tests for parallelism and symmetry.

Mathematical Objectives

Students formulate standard mathematical properties of quadrilaterals, that is, characteristics that express relationships between parts of shapes. The sophistication of their analyses increases as they consider shapes in terms of the measurements of their parts. Students further explore properties involving angle and length measurement and discuss parallelism and symmetry. They identify Shape Makers by their properties.

Mac:

📁 Quadrilateral Makers
- 📁 Quadrilateral Exploration 4
  - ◈ Lines and Angles
  - ◈ M Quadrilateral Maker
  - ◈ M Kite Maker
  - ◈ M Parallelogram Maker
  - ◈ M Rectangle Maker
  - ◈ M Rhombus Maker
  - ◈ M Square Maker
  - ◈ M Trapezoid Maker
- 📁 Demonstration sketches
  - 📁 for Quadrilaterals
    - ◈ Parallel Lines and Angles
    - ◈ M Angle Maker
    - ◈ Parallel Segments Maker
    - ◈ Symmetry demo 1
    - ◈ Symmetry demo 2

Windows:

📁 QuadMkrs
- 📁 QuadExp4
  - ◈ Line_Ang.gsp
  - ◈ M_Quad.gsp
  - ◈ M_Kite.gsp
  - ◈ M_Para.gsp
  - ◈ M_Rect.gsp
  - ◈ M_Rhom.gsp
  - ◈ M_Square.gsp
  - ◈ M_Trap.gsp
- 📁 Demos
  - 📁 Quads
    - ◈ Par_Line.gsp
    - ◈ M_Angle.gsp
    - ◈ Par_seg.gsp
    - ◈ Symdemo1.gsp
    - ◈ Symdemo2.gsp

Session	Student Sheet	SS#	Geometer's Sketchpad sketch
1			M Quadrilateral Maker Parallel Segments Maker Symmetry demo 1 Symmetry demo 2 (Mac) M_Quad.gsp Par_seg.gsp Symdemo1.gsp Symdemo2.gsp (Windows)
2–4	Which Shape Makers Can You Use?	12–14	All M Shape Makers
5	How Are They the Same?	15, 16	All M Shape Makers
6 and 7	Shape Maker Riddles: Who Am I?	17–21	All M Shape Makers
8	Angles in Intersecting and Parallel Lines	22–24	Lines and Angles (Mac) Line_Ang.gsp (Windows) Parallel Lines and Angles Par_Line.gsp

SESSION 1

Introducing Measured Shape Makers to Students

Class Discussion

➡ Explain the operation of the measured Shape Makers to students by using a computer connected to an overhead display or by gathering all students around a single computer.

Each measured Shape Maker sketch has "M" as the first character in its name. There is a separate sketch for each type of Quadrilateral Maker.

➡ Open the sketch M Quadrilateral Maker.

Point out to students that, as in the P&C sketches, there are displays of the lengths of the sides and the measures of the angles of the current shape made by the Shape Maker (see Figure QE 4.1). There are also ways to test for parallelism and symmetry, which you will discuss next.

Length(AB) = 88 pixels   Angle(A) = 83°
Length(BC) = 77 pixels   Angle(B) = 78°
Length(CD) = 63 pixels   Angle(C) = 107°
Length(AD) = 81 pixels   Angle(D) = 92°

▲ Show possible parallels   ▲ Show possible symmetry line
△ Hide possible parallels   △ Hide possible symmetry line

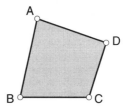

Figure QE 4.1. M Quadrilateral Maker screen

Determining Students' Current Ideas About Parallelism

➡ Have a brief class discussion of the word *parallel*.

I've noticed that some of you are using the word parallel. *What does this word mean?*

Although some students in most classes will have mentioned the word *parallel* prior to its use with the measured Shape Makers, many may not have thought about it yet. It is, therefore, useful to have a short class discussion that deals with how students are thinking about this term. The extended discussion of parallelism in "Teaching Note: What Does *Parallel* Mean?" on page 62 indicates what types of intuitive ideas you may encounter in such a discussion.

➡ Explain to students how mathematicians define parallel lines and segments.

Two lines in a plane are parallel if they do not intersect. Two line segments are parallel if the lines that they determine are parallel.

➡ Illustrate the concept of parallel segments with the sketch Parallel Segments Maker in the Demonstration folder for Quadrilaterals.

As shown in Figure QE 4.2, there are two control points that allow you to change the distance between the lines that pass through the segments, and a control point that allows you to turn the segments. The control points at the segments' endpoints allow you to alter the lengths of the segments. Double clicking on the Show lines button displays the lines through the segments, showing that they are parallel (see Figure QE 4.3).

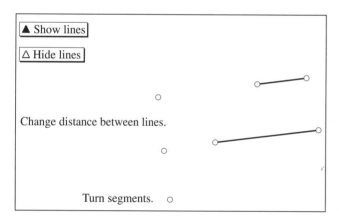

Figure QE 4.2. Parallel Segments Maker sketch display

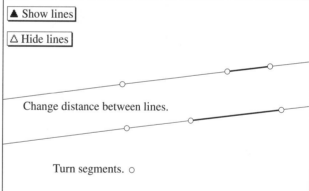

Figure QE 4.3. Double clicking on the Show lines button

Show various configurations of the two segments. Students often think that segments are parallel only if they are the same length, or only if one is directly across from the other. The sketch Parallel Segments Maker can help students see that neither of these conditions is required. Students should now be prepared to understand the parallelism testing tools provided in the measured Shape Makers.

Using the Shape Maker Parallelism Buttons

➡ Open the sketch M Quadrilateral Maker and double click on the Show possible parallels button. Doing so makes four other buttons appear at the bottom of the screen (see Figure QE 4.4).

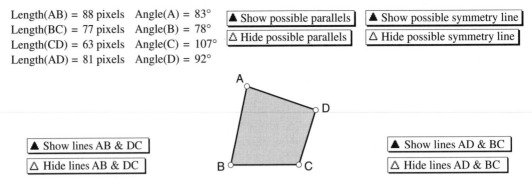

Length(AB) = 88 pixels   Angle(A) = 83°
Length(BC) = 77 pixels   Angle(B) = 78°
Length(CD) = 63 pixels   Angle(C) = 107°
Length(AD) = 81 pixels   Angle(D) = 92°

Figure QE 4.4. Double clicking on the Show possible parallels button displays buttons for showing lines.

➡ Double click on one of the two Show lines buttons that names a pair of opposite sides. This causes the lines containing these sides to appear, with a message appearing if the lines intersect (see Figure QE 4.5).

The scroll bars on the window can be used to view the point of intersection. To hide the lines, double click on the appropriate Hide lines button.

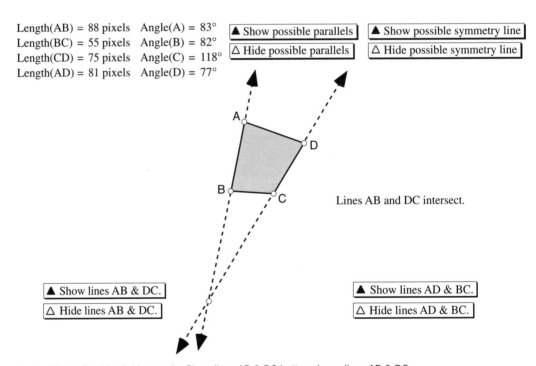

Length(AB) = 88 pixels   Angle(A) = 83°
Length(BC) = 55 pixels   Angle(B) = 82°
Length(CD) = 75 pixels   Angle(C) = 118°
Length(AD) = 81 pixels   Angle(D) = 77°

Lines AB and DC intersect.

Figure QE 4.5. Double clicking on the Show lines AB & DC button shows lines AB & DC.

➡ Ask students how seeing the lines containing opposite sides and the screen message can help them decide whether or not the sides are parallel. Ask students how you should manipulate this Shape Maker so that these two sides become parallel.

How will we know for sure that the sides are parallel? (The "Lines intersect" message will disappear.)

Double click first on the Hide lines AB & CD button, then on the Hide possible parallels button to hide the four parallel lines buttons.

Determining Students' Intuitive Ideas About Symmetry

➡️ Have a class discussion about the word *symmetry*.

Although some students in most classes will have mentioned the word *symmetry* prior to its use with the measured Shape Makers, many may not have thought about it yet. Have students talk about what they mean by the word *symmetry*. See "Teaching Note: Mirror Images" on page 65 for an example discussion that shows how this idea often emerges.

Building on Students' Intuitive Ideas About Symmetry

➡️ To help students build on their intuitive notions of mirror images as described in "Teaching Note: Mirror Images," use the sketch Symmetry demo 1 to discuss the concepts of symmetry and symmetry lines.

(If you demonstrate the sketch with a black-and-white overhead display, be sure to point to objects you're talking about rather than referring to them by color.) See "Mathematical Note: Symmetry" on page 67 for a discussion of the concept of symmetry.

The blue part is the mirror image or reflection of the black part about the dashed (purple) mirror line. You can demonstrate this by double clicking on the Reflect button. (You can change the slope of the dotted line by dragging the red control point on the purple line.) If you consider the shape formed by both the blue and black parts, that shape is symmetric about the dotted line (and the line is called a *line of symmetry*).

You can make various symmetric shapes by dragging the control points on either the blue or black parts. This visually illustrates the mirror nature of the image, because as you move a control point its corresponding control point on the other side of the dashed line moves in exactly the same way—that is, it *mirrors* the movement of its counterpart. You can double click on the Reflect button at any time to illustrate how the black half of the figure flips onto the blue. To see what happens when a whole shape is reflected about a line of symmetry, use the sketch Symmetry demo 2.

Using the Shape Maker Symmetry Buttons

➡️ Open the sketch M Parallelogram Maker and double click on the Show possible symmetry line button. Doing so makes a dashed green line appear. This line can be rotated to any position by dragging its red control point (the one outside the parallelogram).

➡️ Demonstrate how to test a figure for symmetry lines.

Ask students what position you should rotate the green line to so that it becomes a line of symmetry for the parallelogram. After rotating the line to this position, double click on the Symmetry Test button. This causes a blue replica of the original figure to be reflected about the green line (see Figure QE 4.6).

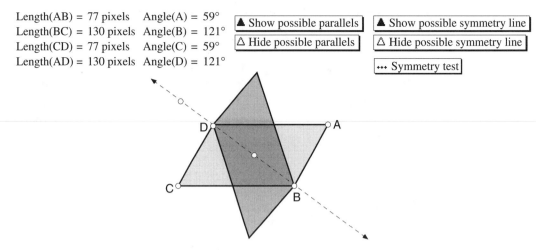

Length(AB) = 77 pixels   Angle(A) = 59°
Length(BC) = 130 pixels   Angle(B) = 121°
Length(CD) = 77 pixels   Angle(C) = 59°
Length(AD) = 130 pixels   Angle(D) = 121°

▲ Show possible parallels   ▲ Show possible symmetry line
△ Hide possible parallels   △ Hide possible symmetry line

••• Symmetry test

Figure QE 4.6. Double clicking on the Show possible symmetry line button shows the dotted line that can be rotated to possible positions for a line of symmetry. Double clicking on the Symmetry test button reflects the figure about the dotted line.

Ask students whether or not the line in the present position is a line of symmetry for parallelogram ABCD. If the blue replica flips exactly onto the original figure, the green line is a line of symmetry; otherwise, it is not. The blue replica will disappear after 10 seconds or when you click the mouse in any blank space on the screen.

Note that sometimes it might not be possible to get the reflected figure to flip onto the original exactly, even though it should. This is due to the fact that Sketchpad must represent the infinite number of points in the plane with the finite number of pixels on the computer screen. Thus, the location of actual mathematical points in the plane can only be approximated by pixel locations.

SESSIONS 2–4

Activity: **Which Shape Makers Can You Use?**

Use Student Sheets 12–14, and sketches:

Mac: Windows:

📁 Quadrilateral Exploration 4   📁 QuadExp4

   ◆ M Square Maker, etc.   ◆ M_Square.gsp, etc.

Students Work in Pairs

➡ Distribute the Which Shape Makers Can You Use? student sheets to the class.

Students are to predict, then check with the measured Shape Makers, which of the shapes described by the student sheet can be made by each Shape Maker. But rather than being shown pictures of the shapes as in Session 1, students are given descriptions of the shapes in terms of their measurements.

If a shape cannot be made by a particular Shape Maker, students explain why not. This encourages them to further reflect on and refine their conceptions of the various Quadrilateral Makers. For instance, we are hoping that students' reasoning is moving away from vague ideas such as "You just can't make different angles with the Parallelogram Maker" toward more property-based reasoning such as "The Parallelogram Maker always has the angles across from each other [opposite angles] equal." These activities encourage students to conceptualize shapes in terms of their properties and to develop ways to communicate about these conceptualizations.

Class Discussion

➡ Have students present their answers and reasons from the Which Shape Makers Can You Use? student sheets.

Students should discuss any disagreements. Ask questions that elicit multiple responses. For example,

> *Mary said that the Parallelogram Maker can't make a polygon with sides of length 20, 30, 40, 50 because that kind of polygon is not a parallelogram— did anybody have a different reason?*

If a student proposes that the Parallelogram Maker won't work because it can't have all different side lengths, ask why.

> *What do we know about the Parallelogram Maker's side lengths that tells us that they can't have these four measurements?*

Generally, there will be some students in class who will say that it's because the Parallelogram Maker must have opposite sides "equal." (Technically, the opposite sides are *congruent*, or have equal length. To be equal, they would have to consist of the exact same set of points. But this distinction is probably too subtle to deal with here.)

Activity: Conjectures and Queries

Class Discussion

➡ In a whole-class discussion, ask students for their conjectures and queries about the Shape Makers: What have they discovered? What are they wondering about?

As students offer their conjectures, other students can comment on, question, or support the conjectures. Encourage the use of clear language. See "Teaching Note: Cultivating Clarity" on page 49 (Quadrilateral Exploration 3) for an example. Also see "Mathematical Note: Precision of Measurements" on page 69 for information about how the precision with which measurements are displayed can sometimes derail productive student conjectures.

Homework/Assessment

➡️ Distribute the How Are They the Same? student sheets to the class.

Students are to describe everything that is the same about all shapes made with each type of Shape Maker. For example, they are to describe everything that all shapes that can be made by the Parallelogram Maker have in common. (They might, for example, say that all such shapes are parallelograms, have opposite sides congruent and parallel, and so on.)

SESSION 5

Activity: How Are They the Same?

Use Student Sheets 15 and 16, and sketches:

Mac: Windows:

📁 Quadrilateral Exploration 4   📁 QuadExp4

　　🔶 M Square Maker, etc.   　　🔶 M_Square.gsp, etc.

Students Work in Pairs

➡️ Have students use the measured Shape Makers to confirm or refute the answers they gave for homework on the How Are They the Same? student sheets.

Students should not erase anything they have written but instead should tell how they would change it or should simply write new answers. They should also write their corrections in a color different from that of their original answers so you can see what each student predicted and how their predictions were corrected when they used the Shape Makers. Collecting these sheets will give you a good idea about where students are in their thinking about classes of shapes. See "Teaching Note: Assessing Students' Ideas About the Student Sheet How Are They the Same?" on page 71.

In this activity, students attempt to determine what is the same about all shapes made with each type of Shape Maker. The activity encourages them to think explicitly about the properties of *classes* of shapes, not particular shapes or Shape Makers. As was mentioned in the introduction, the transition from thinking about Shape Makers and examples of shapes to thinking about a whole class of shapes may be difficult for some students. As was also mentioned in the introduction, it is especially important to be precise with language, clearly distinguishing between Shape Makers and the shapes they make.

Class Discussion

➡️ Have students discuss each problem. As before, students should comment on, question, or support other students' answers.

Activity: Shape Maker Riddles

Use Student Sheets 17–21, and sketches:

Mac: Windows:

   📁 Quadrilateral Exploration 4   📁 QuadExp4

      ◈ M Square Maker, etc.   ◈ M_Square.gsp, etc.

Students Work in Pairs

➡ Distribute the Shape Maker Riddles: Who Am I? student sheets to the class.

Working in pairs, students determine which Shape Maker is described by each riddle. They first predict which *one* Shape Maker is described by a riddle, then describe why they think their prediction is correct. Next, they check their answer with the mea-sured Shape Makers, then write a "proof" that their check is correct. Explain to students that to prove their answers means to give an argument that will convince others that they are correct.

You will probably see several different types of student strategies used in solving these riddles. Some students start by considering all seven of the quadrilateral Shape Makers and proceed by using clues to successively eliminate Shape Makers. For instance, for riddle 4, they might decide that the Parallelogram, Rhombus, Rectangle, and Square Makers are the only Shape Makers that satisfy clue 1. They might elimi-nate the Parallelogram and Rectangle Makers after reading clue 2, and so on. Other students might take each Shape Maker in turn, looking for the one that satisfies all of the clues. Still other students might solve the riddles less systematically. For instance, for riddle 4, a student might say that only the Square and Rhombus Makers are always a parallelogram and kite, but clue 3 eliminates the Square Maker, so the answer must be the Rhombus Maker. All of these strategies are good, but all can be correctly or incorrectly applied.

Class Discussion

➡ Have students discuss their answers as a whole class.

As students present their answers and proofs, other students should challenge any answers and proofs they disagree with. Ask if any students have alternate proofs. Ask questions that encourage critical analysis of solutions:

> *Is that the only Shape Maker that fits the description?*
> *How do you know?*
> *How do you know it's not the Parallelogram Maker?*
> *What other Shape Makers have at least one pair of parallel sides?*

➡ Pay special attention to symmetry.

This activity provides you with an opportunity to check students' ideas about the concept of symmetry of quadrilaterals. Use a computer connected to an overhead dis-play or gather all students around a single computer, and have students discuss their ideas about the symmetry of some of the quadrilaterals. For instance, for riddle 5, some students will think that the only quadrilateral Shape Makers that always have at least two lines of symmetry are the Rectangle and Square Makers. Other students will see that the Rhombus Maker also always has two lines of symmetry. Have students discuss this discrepancy by using the Symmetry test button in the sketch M Rhombus Maker. You can also ask students how, for instance, they know that the Parallelogram Maker doesn't always have at least two lines of symmetry.

Achieving Consensus on Language

Some of the difficulties students have stem from their interpretation of certain terms used in the riddles. For instance, the term *at least* gives some students difficulty. To say that a figure has *at least* two lines of symmetry means that the figure has two or more lines of symmetry.

Disagreements about the meanings of terms can often be resolved by referring to a dictionary. Other times, however, the problem seems to be one of logic and mathematical convention. For instance, some students will object to saying that the Square Maker always has at least two lines of symmetry because, in fact, it always has four lines of symmetry. But, because having four lines of symmetry means that there are two or more, in mathematics such a statement is considered true. Part of students' resistance to such statements often seems to come from their feeling that these statements are not as precise as they should be. For example, saying that the Square Maker has at least two symmetry lines does not seem as precise as saying it has four.

SESSION 8

Activity: Angles in Intersecting and Parallel Lines

Use Student Sheets 22–24, and sketches:

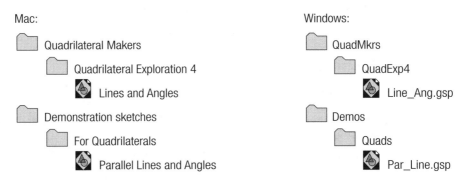

Mac:

Quadrilateral Makers
    Quadrilateral Exploration 4
        Lines and Angles
Demonstration sketches
    For Quadrilaterals
        Parallel Lines and Angles

Windows:

QuadMkrs
    QuadExp4
        Line_Ang.gsp
Demos
    Quads
        Par_Line.gsp

Students Work in Pairs

➡ Distribute the Angles in Intersecting and Parallel Lines student sheets to the class.

Students manipulate several intersecting lines and investigate relationships between the angles formed (see Figure QE 4.7).

In problem 1, students discover that, when two lines intersect, pairs of adjacent angles are supplementary; e.g., m∠A + m∠B = 180°. In problem 2, they discover that, when two lines intersect, pairs of nonadjacent angles are congruent (these angles are sometimes called *vertical* angles); e.g., m∠A = m∠C, m∠B = m∠D. In problem 3, students discover that, when two lines are both intersected by a third line—sometimes called a *transversal*—corresponding angles are congruent; e.g., m∠A = m∠E, m∠B = m∠F. They should also see that alternate interior angles are congruent; e.g., m∠C = m∠E. Also, be sure students see that same-side interior angles are supplementary (e.g., m∠C + m∠F = 180°), because this relationship is useful for judging whether opposite sides of quadrilaterals are parallel. Finally, in problem 4, students use the knowledge gained in problems 1–3 to find all the missing angle measures.

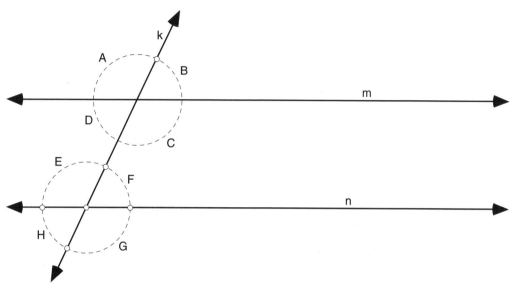

Figure QE 4.7

Class Discussion

➡ Have students discuss their answers and reasons as a whole class.

For problem 1, students might justify that angles A and B are supplementary by arguing that a line (seen as an angle) is "half of a whole rotation." Because a whole rotation has, by definition, 360°, the line must have 180°.

For problem 2, equalities such as m∠A = m∠C might seem visually obvious. Justifying these equalities, though, requires a certain subtlety that some students may not be ready for. For instance, to explain why m∠A = m∠C, students could argue as follows: Angles A and B are supplementary because they form line *m*. But angles B and C are supplementary because they form line *k*. So, since m∠A + m∠B = 180° and m∠B + m∠C = 180°, it must be true that m∠A = m∠C.

For problem 3, probably the most accessible way for students to understand that corresponding angles are congruent is for them to move line *n* onto line *m* by dragging the red point at the intersection of lines *n* and *k* to the intersection of lines *m* and *k*.

➡ After students have established the three relationships above, ask questions about other relationships.

For instance, "How are angles D and E related?" Students could argue that they are supplementary because (1) A and E are corresponding angles, so they are congruent, and (2) D is supplementary to A, so it must be supplementary to E. (Of course, some students may not completely understand such reasoning.)

Problem 4 probes students' understanding of their findings from problems 1–3. It assesses whether they have really understood the spatial relationships of the angles involved when two parallel lines are intersected by a transversal. (If students are having difficulty getting started on this problem, you can give them the hint to use their findings from problem 3.)

➡ To discuss problem 4, project the sketch Parallel Lines and Angles on a large screen.

The sketch shows the measures of all the relevant angles as the slope of the two parallel lines is varied. You can even repeat problem 4 with a different measure for angle 1. Double click on the Hide other angles button so that only the measure of angle 1 is shown. Ask students to predict what the other angle measures are. Double click on the Show other angles button for students to check their predictions.

TEACHING NOTE

What Does *Parallel* Mean?

Teacher: I've heard some of you use the word *parallel*. Tom, how did you use this word?

Tom: Like, in a parallelogram, the top two sides and the other two sides are parallel to each other.

Teacher: I'm wondering what that word means to all of you. When you think of parallel, Tom, what do you think of?

Tom: Two lines that are almost beside each other but one's a little taller. One's a little farther up.

Teacher: Okay. Do you want to add something to that, Steve?

Steve: I think it's sorta like two lines [holds both hands in a vertical position across from each other] that are like across from each other and they'll never meet if they keep going.

Teacher: Two lines that are across from each other and never meet?

Steve: It's just like that [holds up two fingers parallel and in a vertical position] or like that [holds two fingers parallel and in a horizontal position] or any way. And, if you stretched them [moves two vertical parallel fingers away from each other], the two endpoints would never meet.

Teacher: Okay.

Evan: I agree with him. It's like train tracks [holds his hands parallel]. They keep on going in a straight line. If they stay in that straight line, they'll never intersect and they're parallel with each other.

Jacky: I'm still not clear on what these people mean.

Lindsay: Well, those two tracks, they just keep going straight and they never touch each other [holds her hands parallel]. But they stay like side by side.

Mike: It's like taking two pencils that are the same length and putting them right on top of each other [illustrating with two pencils as shown in Illustration A].

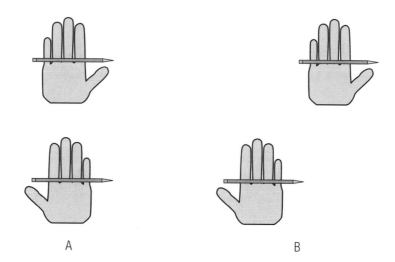

A   B

It wouldn't be parallel if one was not on top of the other [moves one pencil so it is not directly over the other, as in Illustration B].

Teacher: So parallel is like this [holding two different-length rulers in horizontal positions, one on top of the other, with some space in between].

Mike: Yeah, except, except the same length.

Teacher: Do they have to be the exact same length?

Mike: Yeah!

Teacher: What do you think of when you think of parallel lines? I'd like to hear from somebody else.

Juan: I think of lines that are exactly opposite from each other [holds his hands vertically, parallel, and away from each other]. They're the same length exactly. All they are is one line that's on one side, and then there's a big open space, and there's a line on the other side.

Teacher: Do they ever intersect?

Juan: No, they don't.

Li Chen: When Naomi and I were playing the [Shape Maker Challenge] game, she made a shape and we were thinking of why I couldn't make it. We were sorta stuck on if two of the lines were parallel or not. So, I was wondering, do they have to be the same length?

Teacher: Good question. So you're wondering if we could call these two line segments parallel even though they're not the same length [holding up two rulers of different length, horizontal and parallel].

Teacher: Toshi, I haven't heard from you yet. What do you think about this?

Toshi: Well, it might be. 'Cause like when you're skiing, sometimes when you turn, one ski goes farther than the other. And, they still call it parallel skiing [puts both hands next to each other].

Teacher: That one ski [moves the smaller ruler to the left] may be in front of the other ski, but I'm still keeping them—

Toshi: Parallel!

Teacher: They are parallel to each other regardless of how I'm going down that hill. [Rotates the pair of rulers back and forth, but keeps them parallel as in skiing. Toshi nods his head yes.]

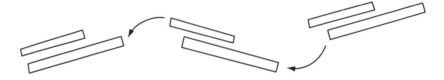

Keisha.

Keisha: Like in a parallelogram, the lines [top and bottom sides] are always even and so are the two lines that connect both of the lines [holds out a finger from each hand]. So, I think that they always have to be even.

Teacher: Okay! Did you hear her argument? Keisha said in a parallelogram these two sides are even [holds her hands vertically] and the other two sides are even length [holds her hands horizontally]. And so Keisha thinks that to be parallel the sides have to be even lengths.

Brenda: Well, I think that maybe they have to be even lengths. But they can be like, one of them could be ahead of the other one [holds her right hand above and parallel to her left hand and then moves her right hand over to the right]. Like I think with the Parallelogram Maker you can like make it kinda at a slant. And, the top line is like farther ahead than the other one. But they're still the same lengths.

Jocelyn: I have a question. If you were going forever, you said that the parallel lines couldn't intersect. If you had these two sides that were right there and they didn't intersect, would they still be parallel without going on forever? Just how would you know if they were still parallel? 'Cause I thought—I might have misunderstood—but I thought you said if they didn't go forever they weren't parallel.

Teacher: Okay, I'm trying to understand exactly what you are asking. [Holding up two rulers of the same length] Are these parallel?

Jocelyn: Yeah!

Teacher: In your mind they're parallel because they're the same length?

[Jocelyn nods her head yes.]

Let's go back and answer the question. [Drawing two different length segments] Are these two line segments parallel even though they're not the same length? I'll give you a minute to think about this.

How many of you say that these segments are parallel? [14 students raise their hands.] How many say they are not parallel? [10 students raise their hands.] Okay, Juan, what did you say, and why?

Juan: I said yes because if they wouldn't be parallel they'd have to intersect. Because if they're parallel they'd be across from each other.

Li Chen: I think that the two lines aren't parallel because they aren't the same length. I thought that because, if you think of a parallelogram, the two lines across from each other are the same length. So, I think that's why they call it a parallelogram, because if it was like that [pointing to the two segments that were drawn different lengths] then I don't think it would be a parallelogram.

Teacher: Do you think all shapes that have parallel sides in them are called parallelograms?

[Li Chen nods her head yes.]

Okay. Well, unfortunately, we are out of time. Great discussion. But keep thinking about this idea of parallelism, and we will talk about it more.

[Lindsay stays after class to ask the teacher a question. She wonders if shapes like these at right are parallel.]

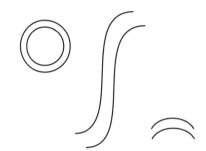

Analysis. This episode nicely illustrates the variety of student conceptions of parallelism. Students have drawn their ideas from several sources—from real-world experiences such as skiing, to manipulations with the Parallelogram Maker, to examination of shapes called parallelograms. The episode also illustrates how some students' concepts of parallelism differ from mathematical conventions. As previously explained, two lines are parallel if they do not intersect, and two line segments are parallel if the lines determined by them do not intersect. Although the students in this episode seem to have the idea that parallel means "staying the same distance apart," some of them also think that parallel segments must be directly across from each other or must have the same length. For the students involved in this class discussion, now is the perfect time to introduce the sketch Parallel Segments Maker. (Note that many students are not explicitly distinguishing lines from line segments. Requiring more precise language in this regard can aid students' concept development.)

TEACHING NOTE

Mirror Images

The students are having a class discussion on the Kite Maker and mirror images.

Geoff: Whenever you would move the Kite Maker, it would be like a mirror.

Evan:	I noticed that too, because let's say in the [Shape Maker Challenge] game that someone got a shape and they made it and you had the Kite Maker and you were trying to stretch the sides. When you would pull a side up, the other side would go out, I think because they [the sides] have to be congruent.
Teacher:	So you noticed that one side and the adjacent side—the side that it touches—that those two sides were congruent to each other. [Evan nods yes.] What does the word *congruent* mean?
Kyle:	Fit exactly [puts one hand on top of the other].
Teacher:	Okay. They fit exactly on top of each other. What does that have to do with being a mirror image?
Geoff:	It's kinda like if you would stand in a mirror and go like this [waves a hand back and forth in a horizontal motion].
Teacher:	And does the Kite Maker always have a mirror image?
Geoff:	Most of the time.
Teacher:	Has anybody else noticed anything about the Kite Maker?
Lindsay:	Well, it's sorta like the same thing. It's just like when you moved one control point [motions with both hands for control points] the other one came down so it would stay congruent. If you moved one control point, the other control point, the opposite control point, would move with it [moves hands while talking]. So it would stay in like a mirror. Like, if you went to a mirror and you took your hand up [raises her right hand in the air] the other hand would move. What you saw in the mirror right across from it would move up.
Teacher:	And is that what you were talking about, Geoff? [Geoff nods yes.] So you noticed that one control point caused one of the line lengths to get longer, so the other one mirrored it and did the same thing.
Carolyn:	I think the Quadrilateral Maker is not exactly like the kite because it doesn't have to be like a mirror. And it doesn't have to have, like, congruent sides or anything.
Teacher:	So are you saying some of the shapes aren't moving in mirrored images? [Carolyn nods yes.] Do you think that might be a characteristic of shapes, then, that they do or don't move in mirrored images? [Students nod yes.] That might be something you want to look at. Geoff, did you notice that any other shapes moved in a mirrored image?
Geoff:	I just basically looked at the kite.
Lindsay:	Well, for the Rhombus Maker, if you were to move one of the control points, the other control point would like sway with it [moves both hands back and forth together in a swaying motion]. Like if, picture like a box in your mind [uses two fingers from her right hand and two fingers from her left hand to form a rhombus]. If you move the right bottom corner [motions as if she is moving the corner], the left bottom corner [motions with her left hand] would move in the same direction [moves her left and right hands together in a swaying motion] with it.
Teacher:	I like the way you gave us some visual imagery to think about; thank you.
Carolyn:	And also, they have to have even sides. So if you try to make them not have even sides [motions with her hand], then the whole shape will move almost.

Teacher: So I hear you starting to look at the parts of the shapes.

Linnea: If you talk about the mirror, it's sorta like that mirror game you play with two people, where you, like, have to move [moves her hand in an arc] however they move.

Teacher: Naomi and Linnea, could you come up here and show us what you mean? [The girls demonstrate.] Geoff, is that what you were talking about with the mirrored image that you noticed with the Kite Maker?

Geoff: Yeah!

Carolyn: Because whenever you look in a mirror, it always looks like you're looking at the opposite thing. That's why the opposite side moves [moves both hands outward], because they have to stay even.

Teacher: You have some nice thoughts that I'd like you to continue to investigate the next time you're on the computer. Think about what other people have said, try them on the computer, and see what you think about this mirror idea.

Analysis. These students have some very clear ideas and vivid imagery concerning how the Kite Maker moves. Many of the students have seen how moving a control point on the Kite Maker (specifically, points B and D) has the same effect as moving one of your hands while looking in a mirror. The reflected hand does the same thing—that is, *mirrors*—what the real hand does, just as the opposite control point on the Kite Maker mirrors what you do with the manipulated control point. Furthermore, some students are inferring that this mirror action is due to the fact that certain adjacent sides must remain congruent.

But even though the students' ideas might suggest to us that they are thinking about symmetry, their ideas might not match exactly the formal concept of mathematical symmetry. So our goal is to guide the development of students' "mirror" ideas so that they become more formal. We do this, first, by having students clearly elaborate their own intuitive ideas, as was done in this class discussion, and, second, by helping students see how their intuitive ideas are related to a more formal conception, as is done with the symmetry demonstration sketches.

MATHEMATICAL NOTE

Symmetry

Some students will intuitively start characterizing shapes by referring to symmetry. Their ideas about this concept may be vague or not completely correct. This note discusses the mathematical concept of symmetry.

The most common form of symmetry, and the one most relevant in our study of quadrilaterals and triangles, is line symmetry. There are many ways to define this type of symmetry. In the elementary grades, students usually learn that a figure is symmetric with respect to a line if half the figure, together with the reflection of that half figure about the line, is equal to the original figure.

Line symmetry is also called *mirror symmetry,* and the line of symmetry is sometimes called a mirror line. To illustrate, the dashed line in Figure QE 4.8 is a mirror line or line of symmetry because, when a mirror is placed along the line perpendicular to the plane of the figure, the half of the shape still visible, together with its mirror image, is identical to the original whole figure (see Figure QE 4.9).

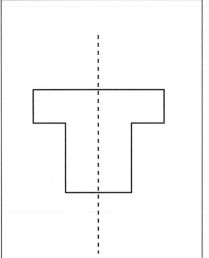

Figure QE 4.8. The original figure

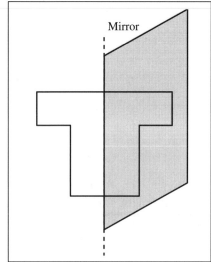

Figure QE 4.9. Half of the original figure and its reflection in the mirror

Another way to define symmetry is in terms of reflections about a line. A figure is symmetric about a line if the mathematical reflection, or flip, of the figure about that line is equal to the figure (that is, lands right on top of the original figure). For instance, consider the shape shown in Figure QE 4.10. As shown in Figure QE 4.11, according to our first definition of symmetry the dashed line is not a line of symmetry. The left half of the shape, together with its reflection in the mirror, does not look like the original shape—so the line is not a symmetry line. We can draw the same conclusion by examining the shape and its reflection about the dashed line. As we can see in Figure QE 4.12, the reflection of the shape does not coincide with the original shape.

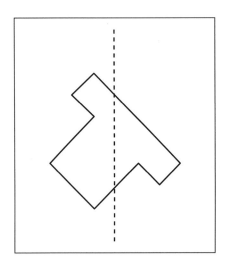

Figure QE 4.10. The original figure

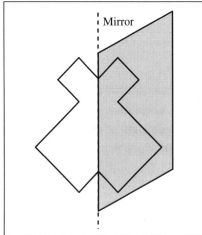

Figure QE 4.11. Part of the original figure and its reflection in the mirror

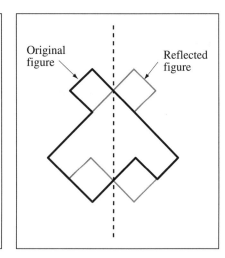

Figure QE 4.12. The original figure and its reflection about the dotted line

Precision of Measurements

Although the computer calculations utilized in Sketchpad are extremely accurate, the precision with which they are displayed is limited. In particular, displayed measurements in the *Shape Makers* sketches are rounded to the nearest whole unit. Most of the time, this is not a problem. But sometimes the limited precision of the display will cause computations with displayed measurements to be slightly inconsistent with theoretical mathematical results.

Example 1. Suppose that the actual measures of the angles of a polygon are 91.4°, 89.9°, 89.4°, and 89.3°. Their sum is exactly 360°. However, because displayed measurements in the *Shape Makers* software are rounded to the nearest whole unit, these angles would be displayed as 91°, 90°, 89°, and 89°, respectively, so the sum of the displayed angles would be 359°.

Example 2. Angle measurements rounded to the nearest unit suggest that the triangle shown in Figure QE 4.13 is equilateral, so all of its sides should be congruent. But, according to the displayed measurements, the side lengths are not all equal. However, examining the angle measurements rounded to the nearest hundredth, as shown in Figure QE 4.14, indicates that the angles are not really congruent, so the sides should not be congruent.

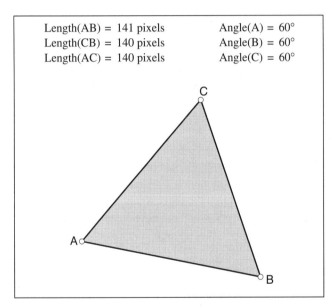

Length(AB) = 141 pixels   Angle(A) = 60°
Length(CB) = 140 pixels   Angle(B) = 60°
Length(AC) = 140 pixels   Angle(C) = 60°

Figure QE 4.13

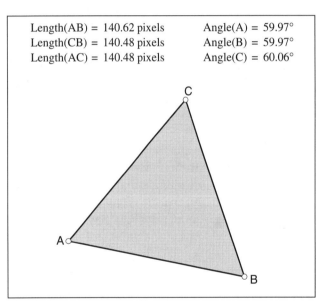

Length(AB) = 140.62 pixels   Angle(A) = 59.97°
Length(CB) = 140.48 pixels   Angle(B) = 59.97°
Length(AC) = 140.48 pixels   Angle(C) = 60.06°

Figure QE 4.14

Analysis. The discrepancy between displayed and actual measurements can confuse students. For instance, if students find that most of the time the sum of the angles of a convex quadrilateral is 360° but sometimes it is not, they might not make the correct generalization. If you see this happening, demonstrate to students that displayed measurements are being rounded. You could, for instance, show them example 1. Then show them how to change the precision of displayed measurements (discussed below) and have them reinvestigate their angle-sum conjecture.

As another example of how changing the precision of displayed measurements can help students, consider a student who found a right triangle in which all displayed side lengths were 2 pixels, contradicting her conjecture that the side opposite the right angle always had to be longest. When the teacher had her change the precision of measurements to tenths, the student saw that her conjecture was, in fact, true for this triangle.

Changing the Precision of Measurements in the *Shape Makers* Sketches. To change the precision of displayed angle or length measurements, choose **Preferences** from the **Display** menu. Then in the dialog box set the Distance Unit or Angle Unit to the desired precision (see Figure QE 4.15).

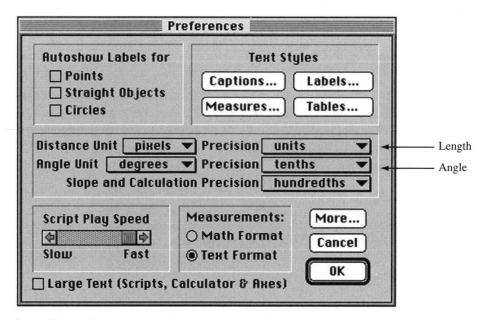

Figure QE 4.15. The Preferences dialog box (Macintosh).

Sometimes it is useful to have Sketchpad compute operations on measurements by using **Calculate** in the **Measure** menu. In this case, the extremely accurate measurements stored by the program are combined, then that very accurate result is rounded. Refer to *The Geometer's Sketchpad User Guide and Reference Manual* for details on how to have Sketchpad make calculations with measurements.

Assessing Students' Ideas About the Student Sheet How Are They the Same?

You will observe a variety of responses and levels of sophistication in students' descriptions of how shapes made by a given Shape Maker are the same.

Level 1. Students give no properties or only incorrect properties of classes of shapes. Examples:

[Square Maker]	*It is always a square.*
[Rectangle Maker]	*It is always a rectangle unless it makes a square.*
[Square Maker]	*They all have to be squares or diamonds.*
[Trapezoid Maker]	*The whole shape has uneven sides, right angles. It looks like this [has drawing].*
[Parallelogram Maker]	*They have uneven sides, right angles 90°, and a mirror image.*
[Parallelogram Maker]	*They have mostly a tilt in their shape. It can make a lot of other shapes than just a parallelogram.*

Level 2. Students give some properties that are necessarily true of the shapes but do not give a set of properties sufficient to characterize the class. Examples:

[Parallelogram Maker]	*They all have at least one set of parallel lines.*
[Rectangle Maker]	*They all have 90° angles and they all have right angles.*
[Square Maker]	*The shapes always have four sides that are equal to each other [has a correct drawing].*
[Rectangle Maker]	*The shapes always have two long sides that are parallel to each other and two shorter sides that are parallel to each other [has a correct drawing].*
[Parallelogram Maker]	*It always has two parallel lines.*
[Kite Maker]	*They all have a line of symmetry. [The student would have to specify that the line of symmetry is through opposite vertices for this property to be sufficient.]*
[Kite Maker]	*Two different pairs of lines, each pair with different length.*

Level 3. Students give properties that are sufficient to characterize the class. Examples:

[Rectangle Maker]	*It always has to have four right angles.*
[Square Maker]	*1. They all have four right angles. 2. They all have four even sides. 3. They are all squares.*
[Rectangle Maker]	*They all have 90° (right) angles, and opposite sides are equal. Also, opposite sides are parallel. They are quadrilaterals too. They always have four sides and four vertices.*
[Rectangle Maker]	*All of the shapes made by the Rectangle Maker have four sides, the corners all make right angles, and the opposite sides are parallel.*

[Parallelogram Maker]	*They all have opposite sides of equal length and opposite sides parallel.*
[Parallelogram Maker]	*It always has two sets of parallel lines.*
[Parallelogram Maker]	*The top and bottom sides are parallel and the left and right are parallel and the same length. If one line (side) is slanted, the line opposite it is slanted; it's the same if it's straight.*
[Rhombus Maker]	*1. They all have a line of symmetry. 2. They all have equal-length sides.*
[Kite Maker]	*It always has to have a line of symmetry from one vertex to another vertex.*

Sometimes it's difficult to know whether students are having difficulty expressing their ideas or their concepts are fuzzy. For example, about the Parallelogram Maker, William said, "The shape has to be parallel no matter how you made it." It is not clear exactly what he meant by *parallel*. He may have understood the concept correctly and simply described it vaguely. Or he may not have really understood the idea that both sets of opposite sides must be parallel.

Other times, students state a property less precisely than they probably intend. For example, about the Kite Maker, Elaine said, "They all have a line of symmetry." She probably meant that the line of symmetry would run vertex to opposite vertex (as had been discussed in class previously), but she did not say so. She may not have clarified her statement because it didn't occur to her at the time that some quadrilaterals have a line of symmetry not passing through two of their vertices.

Shape Maker Mysteries

Summary

In the context of a series of "whodunnit" mystery stories, students are to discover which unlabeled Shape Maker "committed a theft." They are given two types of clues to help them solve the mysteries: (1) they are shown unlabeled pictures of Shape Makers, and (2) they are given verbal statements related to properties of the Shape Makers.

Mathematical Objectives

These activities are intended to sharpen students' knowledge of the properties of the Shape Makers, as well as encourage the type of logical deduction often required in mathematics.

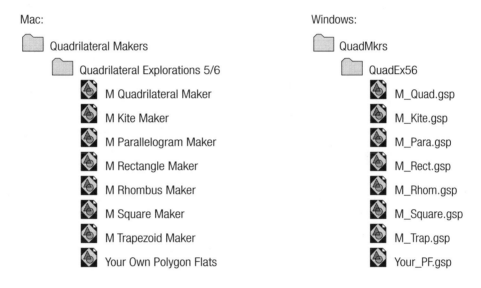

Mac:

📁 Quadrilateral Makers

   📁 Quadrilateral Explorations 5/6

     ◈ M Quadrilateral Maker

     ◈ M Kite Maker

     ◈ M Parallelogram Maker

     ◈ M Rectangle Maker

     ◈ M Rhombus Maker

     ◈ M Square Maker

     ◈ M Trapezoid Maker

     ◈ Your Own Polygon Flats

Windows:

📁 QuadMkrs

   📁 QuadEx56

     ◈ M_Quad.gsp

     ◈ M_Kite.gsp

     ◈ M_Para.gsp

     ◈ M_Rect.gsp

     ◈ M_Rhom.gsp

     ◈ M_Square.gsp

     ◈ M_Trap.gsp

     ◈ Your_PF.gsp

Required Materials

Session	Student Sheet	SS#	Geometer's Sketchpad sketch
1–3	The Mystery of Polygon Flats #1–4	25–32	All M Shape Makers
4 and 5	Make Your Own Mystery of Polygon Flats	33–36	Your Own Polygon Flats (Mac) Your_PF.gsp (Windows)

Activity: The Mystery of Polygon Flats

Use Student Sheets 25–32, and sketches:

Mac: Windows:

    Quadrilateral Exploration 5/6   QuadEx56

       M Square Maker, etc.   M_Square.gsp, etc.

Class Discussion

➡ Distribute the The Mystery of Polygon Flats #1 student sheet to the class, and explain what the task is:

Famous female detective Shirley Lock-Holmes is in the two-dimensional town of Polygon Flats investigating a theft in Quadrilateral Mansion—someone stole the owner's special cream-filled cupcakes. Seven people live in the mansion. These people can change their shape and size, but only to a shape that can be made by their Shape Maker. They are:

Sudha Square	*(played by the Square Maker)*
Rectangle Rick	*(played by the Rectangle Maker)*
Kaneisha Kite	*(played by the Kite Maker)*
Parallelogram Pete	*(played by the Parallelogram Maker)*
Trapezoid Tracy	*(played by the Trapezoid Maker)*
Ricardo Rhombus	*(played by the Rhombus Maker)*
Quentin Quadrilateral	*(played by the Quadrilateral Maker)*

When the theft occurred, the people in the mansion looked like this in the TV room monitors [point to the graphics on the student sheet]. Your job is to determine who committed the theft. You must prove your answer is correct, because whoever you accuse of the crime will have a good lawyer.

You are given two types of clues: (1) You are shown what the people (Shape Makers) looked like and what rooms they were in at the time of the theft; (2) you are provided important information about people in the mansion. There are several mysteries to solve, each more difficult to prove than the one before.

Note that for most of the mysteries you will not be able to determine the whereabouts of each character. You are given only enough information to determine the identity of the thief.

You can use the Shape Makers on the computers as well as your notes and previous student sheets to help you solve the mysteries. Good luck!

Students must use logical reasoning and their knowledge about the Shape Makers to solve the mysteries; then they must write out a complete argument that proves their answer correct. Example solutions are given later in this Exploration, and illustrations of students' reasoning on these tasks are given in "Teaching Note: Right Answers, Incomplete Reasons" on page 78.

Students Work in Pairs

➡ Have students use the measured Shape Makers on the computer to help them figure out answers for Mystery #1.

Use of the Shape Makers makes the problems accessible—most students who have difficulty thinking logically in abstract contexts seem to be able to make inferences based on their manipulations of the Shape Makers.

➡ As you're interacting with pairs of students, if you see students with incorrect answers, you might challenge them in some way:

Are you sure that the Quadrilateral Maker is the only Shape Maker that can make this figure?

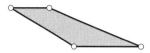

The reasoning required in solving these mysteries is similar to that required in forming geometric proofs. For a completely rigorous solution, students must use logical deduction and justify each step in their arguments. However, keep in mind that many middle-school and junior-high students have difficulty reasoning so abstractly. Do not make students whose arguments are not completely rigorous feel that their answers are incorrect; they are passing through a natural stage in the development of the kind of abstract reasoning we are trying to encourage. Through class discussions of student solutions, students will cooperatively establish standards of rigor that they feel comfortable with. Your task is to encourage them to make their standards as rigorous as they can.

Class Discussion

➡ Have several students present their solutions for Mystery #1, step by step.

Other students should ask questions about statements they don't understand and challenge any statements they think are false. Although there is only one correct answer to who committed the crime, there may be several valid ways to deduce who the thief is. Even so, students may offer incorrect or incomplete solutions like some of those illustrated in "Teaching Note: Right Answers, Incomplete Reasons" on page 78.

In the class discussion, encourage students to challenge each other's arguments. If no other student challenges an incorrect or incomplete argument in the class discussion, you might challenge it, as illustrated in "Teaching Note: Right Answers, Incomplete Reasons."

Students Work in Pairs

➡ Have students use the measured Shape Makers on the computer to help them figure out answers for Mysteries #2, 3, and 4.

When all students have finished Mystery #2, you can have a class discussion on it, then let students go back to working on the other mysteries.

Example Solutions

Mystery #1. Only the Trapezoid Maker and the Quadrilateral Maker can make the shape in the study (so only they could have committed the theft). But the Quadrilateral Maker is the only Shape Maker that can make the shape in the game room. Thus, Trapezoid Tracy must have committed the theft.

Mystery #2. The only Shape Makers that can have four unequal sides are the Quadrilateral Maker and the Trapezoid Maker, so clue 2 indicates that these two are our only possible suspects. Since the theft occurred in the library and the Quadrilateral Maker must be in the parlor (because it is the only Shape Maker that can make the nonsquare shape), Trapezoid Tracy must be the thief.

Mystery #3. The clues in this mystery are less straightforward. Although clue 1 tells us the room the theft occurred in, it does not eliminate any suspects because all the Shape Makers can make a square. Clue 2 means that the Square Maker and the Rectangle Maker are in the same room, which, according to the picture, must be the study. Clue 3 tells us that the Square, Rectangle, or Rhombus Maker was out of the house, because these are the only Shape Makers that always have at least two lines of symmetry. But since we know that the Square and Rectangle Makers were in the house, clue 3 tells us that the Rhombus Maker was out of the house. Of the shapes shown in the picture, the Kite Maker can make only the squares. But the square in the study is the Square Maker. Thus, the square in the parlor must be the Kite Maker; so Kaneisha Kite must be the thief. (You can also reason that the Quadrilateral Maker must be the shape in the game room and that the only Shape Makers left that could make the shapes in the library and living room are the Parallelogram Maker and Trapezoid Maker. Thus, by the process of elimination, the Kite Maker must be in the parlor.)

Mystery #4. One "proof" that Rectangle Rick committed the theft goes as follows: Because the theft occurred in the game room, the thief must be able to make the shape shown in this room. Sudha Square, Kaneisha Kite, and Ricardo Rhombus cannot make this shape. Although Quentin Quadrilateral and Trapezoid Tracy can make this shape, they are eliminated as suspects because they don't always have at least two equal sides (clue 2). Thus, we have eliminated all but two suspects—Parallelogram Pete and Rectangle Rick. To implicate Rick, we must show that Pete is not in the game room. The shape in the parlor could only be made by Kaneisha Kite, Parallelogram Pete, Trapezoid Tracy, Ricardo Rhombus, or Quentin Quadrilateral. Clue 2 eliminates Tracy and Quentin. Clue 3 indicates that Kaneisha Kite was at the airport, so she couldn't be in the parlor. The only Shape Makers that always have at least two lines of symmetry are the Square, Rectangle, and Rhombus Makers, so clue 4 indicates that Ricardo Rhombus is in the study and thus couldn't be in the parlor. That means that Pete is in the parlor. Therefore, Rick is the culprit.

SESSIONS 4 AND 5

Activity: Make Your Own Mystery of Polygon Flats

Use Student Sheets 33–36, and sketch:

Mac:

Quadrilateral Exploration 5/6
Your Own Polygon Flats

Windows:

QuadEx56
Your_PF.gsp

Students Work in Pairs

➡ Have student pairs make their own mysteries for Polygon Flats (see the directions on page 1 of the Make Your Own Mystery of Polygon Flats student sheet).

This is a task that students get quite excited about, but it is also difficult to do correctly.

To show where the Shape Maker personalities were during the theft, students put the Shape Makers in the sketch Your Own Polygon Flats into rooms, then double click on the Hide names button to hide the names of the Shape Makers. They then write their own mystery clues.

➡ After students in each pair have made their mystery, ask them to try to solve it themselves. This helps them check their logic and clues and make appropriate revisions.

➡ Have each student pair give their mystery to another pair to solve.

For instance, Pair 1 should have Pair 2 try to solve Pair 1's mystery. Pair 1 should silently observe Pair 2 try to solve Pair 1's mystery so that Pair 1 students understand any difficulties that Pair 2 has or any errors that Pair 1 might have made in logic or in writing clues.

After Pair 2 has deduced a solution for Pair 1's mystery, they should double click on the Show names button to see if their answer is correct. The two pairs should then talk about any errors that occurred in either the solving or the writing of the mystery. They should then switch roles, with Pair 1 trying to solve Pair 2's mystery.

If students find that they made mistakes in their constructions of mysteries, they should revise them, then test them again by sharing them with another pair of students.

Class Discussion

➡ After all student pairs have traded mysteries, have a class discussion of some of the mysteries and solutions.

There may be disagreements over whether the given clues are sufficient or whether there is more than one solution. Have the class arbitrate these disagreements.

When all students have completed their revisions and are satisfied that they have made good mysteries, the class might put together a book of mysteries. You (or the students) might even want to copy student pictures from the Your Own Polygon Flats sketches into a word processor document containing their clues. (Or you could print these pictures and tape them onto Student Sheet 35.)

Homework

➡ Distribute the Relating Shape Makers student sheets to the class. Students are asked questions that encourage them to think about interrelationships between various Shape Makers. They predict answers for homework, then on the following day test their predictions using the Shape Makers.

If you have students write their predictions in one color and the answers they find in pairs at the computer on the next day in another, you can collect these sheets and use them to assess your students' individual thinking.

Right Answers, Incomplete Reasons

You will find many students who find correct answers to the mysteries but present incorrect or logically incomplete arguments. Students giving incomplete solutions should be encouraged to elaborate their proofs. Questions such as "How did you know that this Shape Maker didn't make this shape?" can be very helpful in encouraging students to further justify their answers.

Students' reasoning can also be flawed because they have mistaken notions about what shapes the Shape Makers can make. For instance, one student thought that the Quadrilateral Maker had to have at least two equal sides—thus, clue 2 in Mystery #4 did not eliminate this Shape Maker as a suspect for this student.

Another way to help students who are stuck is to ask them to tell you what they are thinking: "What does this clue tell you? What Shape Makers could be in this room?"

Several levels of sophistication will be seen in students' solutions, as illustrated here.

Mystery #2, Darla. "I think it's the Quadrilateral Maker because we put all the other Shape Makers in rooms, and clue 2 kind of gives it away." (She listed the Quadrilateral Maker and the Square Maker in the parlor; the Kite Maker in the study; the Rectangle Maker in the den; the Trapezoid Maker in the library; the Rhombus Maker in the living room; and the Parallelogram Maker in the game room.)

This student used her knowledge of the Shape Makers to put each Shape Maker character in a room that had a shape she thought it could make. Although she listed correct possible room locations for the Shape Maker characters, her solution seemed to ignore these locations and focus only on clue 2.

Mystery #3, Darla. "Because of clue 2, the Square Maker and the Rectangle Maker are in the study. Because of clue 3, the Rhombus Maker was out of the mansion. I put the Trapezoid Maker in the library, the Parallelogram Maker in the living room, and the Quadrilateral Maker in the game room because I know they can make these shapes. The Kite Maker is the only person left, so the thief was Kaneisha Kite."

Darla did not carefully consider and eliminate all the possibilities. She did not explain why the Kite Maker could not be in the library, living room, or game room.

Mystery #3, Megan. "Clue 2 means that the Square Maker and the Rectangle Maker are in the study. Clue 3 means that the Rhombus Maker was out of the mansion. So the thief was not the Rectangle Maker, Square Maker, or Rhombus Maker. I remember from another student sheet that the Trapezoid Maker can make the shape in the library. The Kite Maker can't make the shapes in the living room or library, but the Parallelogram Maker can. I remember from another sheet that the Quadrilateral Maker can make the shape in the game room. So the thief can't be the Rectangle Maker, Square Maker, Rhombus Maker, Parallelogram Maker, Trapezoid Maker, or Quadrilateral Maker. So it has to be the Kite Maker."

Megan seems to have most of the pieces for a complete argument. The only thing she did not do is say why the Kite Maker could not make the shapes in the library and living room. Also, the order in which she presented her ideas and the second to last statement suggest that she thought she had eliminated all other possibilities, but she hadn't. For instance, she did not say why the Parallelogram Maker could not be in the parlor.

Mystery #3, Juanita and Carlos. Both students used clues 2 and 3 to conclude that the Square Maker and Rectangle Maker were in the study and the Rhombus Maker was out of the mansion. Thus, the Kite Maker could not have been out of the mansion or in the study, so it had to be in the parlor, library, living room, or game room. But, of the shapes shown in these rooms, the Kite Maker could only make the square and thus was in the parlor. Juanita said that the Kite Maker could not make the shapes in the library, living room, or game room because it had to have two pairs of adjacent sides equal. Carlos justified this same claim by saying that the Kite Maker had to have a line of symmetry "angle to angle" (meaning vertex to vertex), which none of these shapes have. These students devised different, but equally sound, proofs.

Mystery #4, Manuel and Toshi. After reading all the clues, Manuel and Toshi conclude that the person in the study is Ricardo Rhombus because of clue 4. They deduce from clue 3 that Kaneisha Kite was out of the mansion at the time of the theft. From clue 2, they derive two suspects—Rectangle Rick and Parallelogram Pete. They then claim that Rectangle Rick must be in the game room—and be the thief—because he couldn't make the shape in the parlor, so Parallelogram Pete had to be in the parlor.

Manuel and Toshi did not explicitly say why Parallelogram Pete had to be in the parlor (hence, their argument was incomplete). So the teacher questioned them about this conclusion: "As Rectangle Rick's lawyer, I don't see how you can say that Parallelogram Pete had to be in the parlor—it might have been Quentin Quadrilateral or Trapezoid Tracy because both of them can make that shape." Manuel and Toshi countered that neither Quentin Quadrilateral nor Trapezoid Tracy could have been in the parlor because neither of them always has two equal sides, which clue 2 said had to be true about the person in the parlor. Thus, although Manuel and Toshi's argument was incomplete, they had made the correct conclusion and were able to satisfactorily rebut challenges to their conclusion.

Relationships Between Shape Makers

Summary

Students investigate relationships between Shape Makers, as well as relationships between Shape Makers and the classes of shapes they make. Students have debates about inclusion relationships between classes of quadrilaterals, then make a hierarchical classification chart for these classes.

Mathematical Objectives

Students develop an understanding of interrelationships between the classes of quadrilaterals made by each Shape Maker. They develop ways of reasoning needed to formulate and justify these interrelationships.

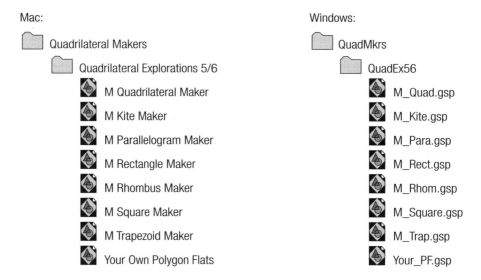

Mac:

📁 Quadrilateral Makers

    📁 Quadrilateral Explorations 5/6

        ◆ M Quadrilateral Maker

        ◆ M Kite Maker

        ◆ M Parallelogram Maker

        ◆ M Rectangle Maker

        ◆ M Rhombus Maker

        ◆ M Square Maker

        ◆ M Trapezoid Maker

        ◆ Your Own Polygon Flats

Windows:

📁 QuadMkrs

    📁 QuadEx56

        ◆ M_Quad.gsp

        ◆ M_Kite.gsp

        ◆ M_Para.gsp

        ◆ M_Rect.gsp

        ◆ M_Rhom.gsp

        ◆ M_Square.gsp

        ◆ M_Trap.gsp

        ◆ Your_PF.gsp

Required Materials

Session	Student Sheet	SS#	Geometer's Sketchpad sketch
1 and 2	Relating Shape Makers	37–40	All M Shape Makers
3 and 4	Mathematical Debates	41, 42	All M Shape Makers
5 and 6	Mathematical Debates (second copy)	41, 42	
	How Are They the Same? (second copy)	15, 16	All M Shape Makers

Activity: **Relating Shape Makers**

Use Student Sheets 37–40, and sketches:

Mac: Windows:

📁 Quadrilateral Exploration 5/6   📁 QuadEx56

◈ M Square Maker, etc.   ◈ M_Square.gsp, etc.

Students Work in Pairs

➡ Have students use the Shape Makers to check their predictions on the Relating Shape Makers student sheets that they completed for homework.

Questions on these sheets have students think about the interrelationships between various pairs of Shape Makers, as well as between Shape Makers and the sets of shapes they make. Such thinking provides the groundwork for students to use the Shape Makers to help them understand relationships between classes of shapes (e.g., squares and rectangles, parallelograms and rectangles). When students see, for example, that a rectangle has all the properties of a parallelogram and also has right angles, they are seeing a relationship that can enable them to see a rectangle as a parallelogram with right angles. This helps them see how one class of shapes is related to another, moving them toward van Hiele level 3.

The first two questions on student sheet 37 encourage students to see that the Rectangle Maker can make every shape the Square Maker can make, but not vice versa. The intent of question 3 is to have students see that the Rectangle Maker can make squares because squares have all the properties of rectangles. Question 4 encourages students to see that if we add the property "All sides must be equal" to the Rectangle Maker, it becomes a Square Maker.

Student sheets 37–40 contain a similar exploration, but one that focuses on the relationship between the Parallelogram Maker and the Rectangle Maker. Obviously, similar explorations could be implemented with other pairs of quadrilateral Shape Makers.

Class Discussion

➡ Have students discuss and justify their answers to the questions on the student sheets.

Students should challenge answers they disagree with. As students explain why they answered the way they did, encourage them to be precise.

You will observe a variety of levels of sophistication in students' reasoning about these questions. For instance, to justify the claim that every shape made by the Square Maker can also be made by the Rectangle Maker (or that the Rectangle Maker can be used to make squares), some students' answers will be more visual and experience-based: "We found before that the Rectangle Maker makes squares and rectangles; it can make squares, just like the Square Maker." Other students' answers might be more analytical and property-based: "The Rectangle Maker makes quadri-laterals with four right angles and opposite sides equal. Because squares have four right angles and all sides are equal, the Rectangle Maker can make them." Finally, some students' answers will be based on an emerging knowledge of interrelationships between classes of shapes: "Because all squares are rectangles."

Being exposed to such diverse justifications will help all students better understand their own thinking. It will also help students who are functioning at the lower levels of sophistication to start moving to higher levels.

Homework

➡ Distribute one copy of the Mathematical Debates student sheet to each student. Students are to decide on the validity of statements about relationships between classes of quadrilaterals, then write arguments justifying their answers.

SESSIONS 3 AND 4

Activity: **Mathematical Debates**

Use Student Sheets 41 and 42, and sketches:

Mac: Windows:

📁 Quadrilateral Exploration 5/6   📁 QuadEx56

◭ M Parallelogram Maker, etc.   ◭ M_Para.gsp, etc.

Students Work in Pairs

➡ Give pairs of students time to talk to each other and to use the Shape Makers to check their answers to the questions on the Mathematical Debates student sheet they completed for homework. They should write all alterations in a color different from the one they used for their homework (so that when you collect these sheets you can distinguish students' individual initial thoughts from those derived in class with the Shape Makers). Tell students to prepare to defend their ideas in a public debate.

Class Discussion—The Debates

➡ When students have completed their student sheets, conduct the mathematical debates, problem by problem.

For problem 1, for example, ask those students who circled "true" to stand on one side of the room and those who circled "false" to stand on the other. Have students on each side take turns giving arguments supporting their answers. Students can switch sides of the room any time they want, but as they do, ask some of them why they changed their minds. (Of course, you don't have to have students stand on opposite sides of the room, but doing so seems to make the activity more exciting for students.)

The questions on the Mathematical Debates student sheet are more abstract than questions dealing with Shape Makers, because the goal is to encourage students to think about *classes* of shapes. To give the usual "mathematically correct" responses, students need to think of each class of shapes as defined by a set of geometric properties. That is, they must think, for example, of a rectangle as a four-sided polygon

---

with four right angles and opposite sides equal, and a square as a four-sided polygon with four right angles and all sides equal. Since all of a square's sides are equal, its opposite sides are equal. Therefore, a square has all the properties of a rectangle, so it must be a type of rectangle.

You will find that many students are unable to perform this type of deductive reasoning and formal classification. Many will instead utilize their knowledge about the Shape Makers to answer these questions. They might reason that rectangles are shapes made by the Rectangle Maker and that squares are shapes made by the Square Maker. But because they have already established that the Rectangle Maker can make any shape made by the Square Maker, they conclude that squares must be rectangles. This reasoning is also acceptable.

It is important to let students argue about and decide on the answers to these questions. But recognize that some of the issues raised are quite difficult for students to resolve—it is common for a class to spend two days of heated debate on these ideas. However, if the activity comes to a close with students still not reaching the "correct" consensus on some of the questions (usually "Every square is a rectangle"), you should let students know what mathematicians have to say about these classifications. Then have students discuss why mathematicians may have chosen to classify these shapes the way they do. Analogies such as the following can help students understand mathematicians' perspective: Think about a Granny Smith apple. Is it an apple? Is it a fruit? See "Teaching Note: Shape Makers as Representations of Classes of Shapes" on page 85 and "Teaching Note: Some Example Debates" on page 86 for descriptions of some correct and incorrect student thoughts about these ideas. Also see "Mathematical Note: The Mathematician's View" on page 91 for a discussion of how mathematicians classify quadrilaterals.

Homework/Assessment

➡ After the debates have concluded, for homework, give students another copy of the Mathematical Debates student sheet and ask them to give the answers they currently believe are true. Collect and examine these sheets to determine each student's thoughts about these classification problems.

SESSIONS 5 AND 6

Activity: Making a Quadrilateral Classification Chart

Class Discussion

➡ Illustrate a hierarchical classification system by giving some examples.

"Things on Earth," for example, can be broken down into the classifications shown here (only some are shown). The arrows indicate that one class of things is a subset of another class.

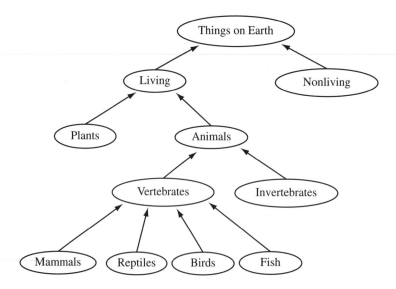

Students Work in Pairs

➡ Have students make a classification chart for the different types of quadrilaterals. "Quadrilaterals" is the overarching category and goes at the top of the chart. Students can use the measured Shape Makers to aid them.

Class Discussion

➡ Have students describe their classification charts and the reasons they organized them the ways they did.

Ask questions about students' charts. For example, looking at the classification chart shown in "Mathematical Note: The Mathematician's View" on page 91, you could ask:

What does this arrow from rhombuses to parallelograms mean?

Why did you say that rhombuses were parallelograms?

Does your chart show that rhombuses are trapezoids? How?

The last question may be difficult for students because it involves the implication "All rhombuses are parallelograms, all parallelograms are trapezoids; therefore, all rhombuses are trapezoids." In fact, some students may want to draw extra arrows showing such relationships. For example, they might add an arrow going from rhombuses to trapezoids.

Assessment

➡ After the class discussion, have students make final versions of their quadrilateral classification charts and hand them in to you along with their original versions. They should also explain why they made any changes that they did.

Activity: Quadrilaterals Assessment

Use Student Sheets 15, 16, 41, and 42.

Students Work Alone

The two major goals of our instruction on quadrilaterals have been for students to progress through the van Hiele levels far enough to (1) think of common quadrilaterals in terms of their geometric properties and (2) be able to use properties to classify quadrilaterals in a hierarchy.

➡ You can assess the first goal by having students individually complete the How Are They the Same? student sheet from Quadrilateral Exploration 4 again (or a similar sheet that you devise).

You can evaluate students' performance on this sheet by comparing their answers to those they gave the first time they completed the sheet and by rereading "Teaching Note: Assessing Students' Ideas About the Student Sheet How Are They the Same?" on page 71 (Quadrilateral Exploration 4).

➡ You can assess the second goal by examining the second copy of students' Mathematical Debates student sheets and their final work on their quadrilaterals classification charts. Do students give correct answers and correct justifications of their answers?

Correct justifications of statements such as "All squares are rectangles" will vary. They can range from arguments that a square has all the properties required of rectangles so it must be a rectangle, to claims that squares are rectangles because squares can be made with the Rectangle Maker.

TEACHING NOTE

Shape Makers as Representations of Classes of Shapes

We want each Shape Maker to become a "concrete" embodiment for the *class* of shapes given by its name. For example, the Rectangle Maker should somehow represent for students the class of all rectangles. The properties embodied by it are exactly those properties that all rectangles have. As instruction continues, you will see most students become increasingly able to use the Shape Makers to help them think about classes of shapes, as these examples illustrate.

Example 1. These students are considering the relationship between squares and rectangles.

Beth: A square is a rectangle, but a rectangle is not a square.

Meg: I agree. The Rectangle Maker can make a square, but the Square Maker cannot make all rectangles.

Sue: Every shape made by the Square Maker can be made by the Rectangle Maker because a square is a rectangle.

Meg and Sue are using mental models involving Shape Makers to understand the relationship between the class of squares and the class of rectangles.

Example 2. To support the claim that every shape made by the Rectangle Maker can be made by the Parallelogram Maker, Jon said: "A rectangle has four right angles and parallel opposite lines, so it has opposite sides parallel and is a parallelogram." Jon had established a clear relationship between the classes of shapes—rectangles and parallelograms—and the corresponding Shape Makers—Rectangle Maker and Parallelogram Maker.

Example 3. To support the claim that a rectangle is a parallelogram but not all parallelograms are rectangles, another student said: "Because you can make other shapes with the Parallelogram Maker."

Analysis. All of these students have used their experiences with Shape Makers to correctly reason about classes of shapes. They seem to see that the class of rectangles is exactly that set of shapes that can be made by the Rectangle Maker, and they can make similar observations about squares and parallelograms. As shown in example 1, from an observation that the Rectangle Maker can make any shape the Square Maker makes, many students conclude that all squares are rectangles. Because these students have firmly established the relationship between the Shape Makers and corresponding classes of shapes as described above, they can reflect on actions with the Shape Makers to draw conclusions about properties of and interrelationships between classes of shapes.

Some students, however, seem to confuse the Rectangle Maker with examples of rectangles. As we discussed in "Teaching Note: Shapes Versus Shape Makers" on page 25 (Quadrilateral Exploration 1), this type of misconception generally persists because the student did not utilize correct language—for example, using the term *rectangle* to refer to both examples of rectangles and the Rectangle Maker.

TEACHING NOTE

Some Example Debates

The excerpts below illustrate students' attempts to make sense of the statements given in the debates—some successful, some not.

Teacher: Let's start with the first statement: Every rectangle is also a parallelogram. [The 25 students have separated into two groups—trues and falses—on opposite sides of the room.] First, I want to hear from the three people who say "false" because maybe their arguments will convince some of you to go to their side.

Amir: Why I think that every rectangle is not a parallelogram is because a rectangle has to have all right angles. But all a parallelogram needs is parallel lines.

Teacher: Raisa, why do you say "false"?

Raisa: Well, because I knew that a parallelogram makes a shape where there are two shorter sides that are slanted.

Note Raisa's use of the word *parallelogram* instead of the more appropriate term Parallelogram Maker. She also uses the vague term *slant,* which causes the teacher to focus the discussion on the meanings of the words students are using.

Teacher: I'm not sure I know what a slant is. How can we define *slant?*

Linnea: If they're more than 90° or less than 90°.

Teacher: What needs to be more than 90° or less than 90°?

Linnea: The angles.

Teacher: What do you think of that idea, Raisa?

Raisa: That's a good way.

Teacher: Okay. So, you're saying that every rectangle is not a parallelogram because . . .

Raisa: A rectangle can only have four 90° angles; a parallelogram can, but it doesn't have to. But the rectangle always has four 90° angles, no matter how you move the control points.

Teacher: Now it's time for the "true" side people to come up with an argument to convince Raisa to move.

Damon: You said that the parallelogram can have four 90° angles. Well, then why couldn't it make every rectangle? Because if it can make four 90° angles and a rectangle always has four 90° angles, why can't it make every rectangle?

Raisa: Give me a minute to think about it.

Teacher: You asked why the parallelogram couldn't make every rectangle. How can a parallelogram make a rectangle?

Damon: Oh, sorry. I meant the Parallelogram Maker could make every rectangle.

Teacher: I thought so. Let's be sure we're clear about what we are talking about, shapes or Shape Makers. Otherwise it can be confusing.

Jocelyn: Can I change to the "true" side?

Teacher: Why do you want to change?

Jocelyn: Well, I was thinking about what Damon said, and I thought if it's a rectangle, why can't it be a parallelogram—if the Parallelogram Maker can make it. Because every rectangle, you could look at it and say it's a parallelogram.

Damon's use of the word *make* indicated that he was talking about a Shape Maker, not a shape or class of shapes. So the teacher took this opportunity to remind the class to distinguish between shapes and Shape Makers. Also, both Damon and Jocelyn thought about their experiences with the Parallelogram Maker to reason about the relationship between rectangles and parallelograms.

Raisa: [Moving to the "true" side of the room] I think it's true, I think. I'm getting confused.

Teacher: That's okay. Sometimes you need that confusion stage before you get to the "aha" stage. Who would like to try to help her?

Kyle: Well, you know how a Shape Maker can only make a shape if it follows the rules. The Square Maker can only make squares. The Parallelogram Maker can only make parallelograms, and it can make a rectangle, so a rectangle has to be a parallelogram.

Lindsay: If a Parallelogram Maker can make a rectangle, do you think a rectangle is considered a parallelogram?

Linnea: Since the rectangle has to have four right angles and two sets of parallel lines, the Parallelogram Maker can make rectangles because it follows those rules. The rectangle is a parallelogram because it follows the same rules. And if a shape follows the same rules, then they can make each other. Then they are each other.

Raisa: Yeah!

Teacher: So, do you think you could also call every rectangle a parallelogram?

Raisa: Yeah!

Teacher: Why?

Raisa: Because, Linnea said if they have the same rules and they can make the shape, then they are each other.

Teacher: Does anybody disagree with what Linnea said? If they can make each other, they are each other. So, question number 1, then, you all say is true, right?

So why don't we go on to question number 2: Every parallelogram is also a rectangle.

[Raisa is the only one on the "true" side.]

Geoff: How did you decide that it is true?

Raisa: Linnea said that if they have the same rules and they can make the shape, then they are each other. So I figured if the first one was true, then why shouldn't this one be true?

Jacky: Raisa, you said first that the parallelogram can slant, right? You said that the rectangle keeps 90° angles, right? It says every parallelogram—that means it can slant—is a rectangle. That means the Rectangle Maker can make a parallelogram. But you said that a rectangle has to have 90° angles.

Raisa: Can you rephrase that?

Lindsay: What I think she's trying to say is that in problem 1, it was having a Parallelogram Maker make a rectangle. Jacky's trying to say that if a parallelogram can have a slant, or not a 90° angle, does a parallelogram have to have a 90° angle at all times?

Raisa: No.

Lindsay: So can a Rectangle Maker make a parallelogram if a Rectangle Maker cannot make less than a 90° angle?

Raisa: No!

Teacher: So, do I hear you saying that maybe every parallelogram isn't a rectangle? [Raisa nods her head yes.]

Let's look at question 3: Is every rhombus also a quadrilateral?

Juan: Yes. Because the Quadrilateral Maker can make shapes that the Rhombus Maker can make.

Li Chen: I said yes because every four-sided shape is a quadrilateral. And a rhombus is a four-sided shape.

Damon: A Quadrilateral Maker can make any four-sided shape. So it can make all rhombuses.

[All students agree that the fourth statement, "Every quadrilateral is also a rhombus," is false.]

Teacher: Question 5: Every square is also a rectangle.

[Linnea, Jocelyn, Manuel, and Raisa are on the "false" side.]

Raisa: I said "false" because when Kimmy and I were exploring the shapes, the Square Maker could not make a rectangle, but the Rectangle Maker could make a square. And we knew that a square could only have four sides that are even to each other and four right angles. But a rectangle, it always has four right angles. It can have four sides that are even to each other, but sometimes it has two longer lines that are parallel to each other and two shorter ones that are parallel to each other.

Teacher: Tell me what you think the question means.

Li Chen: I think it means that you can make a square with the Rectangle Maker. So, you can take the Rectangle Maker and make a square out of it.

Brenda: I think I agree with her. But it's like the Rectangle Maker can make every square that the Square Maker can make.

Jarod: I think it's because the only rules the rectangle has are two sets of parallel lines and four right angles. And every square you make follows those rules.

Adam: [Moving to the "false" side] Even though every square's a square, a square has four even sides, and mainly all rectangles do not.

Linnea: Well, when I read the question I kinda think it's saying like you can make every square into a rectangle. 'Cause it's saying like every square is also a rectangle. And I don't think that that can be true, because every square's rules are four equal sides and four right angles. And every rectangle's rules are four right angles and it has to have two sets [pause]. Well, now I think it's true.

Teacher: Why?

Linnea: I read it over again, and . . .

Teacher: So what do you think the question is really saying Linnea?

Linnea: Can the Rectangle Maker make every single square? [Linnea moves to the "true" side.]

Geoff: I would like to move over there [the "false" side] because a square has four even sides. But then also a rectangle has two even sides and two other even sides. So, I don't think that a square is really a rectangle. Because if it was, it would be two long sides and two short sides.

Teacher: What can I call this, Geoff? [The teacher points to a square she drew on a large sheet of chart paper.]

Geoff: A square.

Teacher: Can I call it anything else?

Geoff: No!

Teacher: This figure right here has all four sides equal and right angles. Geoff said this is called a square and that is the only thing you can call it. Do you agree or disagree, and why?

Rob: A rectangle has four right angles and it has to have two sets of parallel lines, and that is all. But a square has like two sets, so you can call a square a rectangle.

Brenda: I was thinking about the Shape Makers and if they could make each other. When you look at a square, you don't really think of it as a rectangle. But it also follows the rules of a rectangle, so you could call it a rectangle.

Lindsay: I think that the Rectangle Maker can make some shapes, and one of the shapes the Rectangle Maker can make is the square shape, that if you see a square it's a rectangle.

The next day, when the teacher asked the students about questions 5 and 6, almost all students said they thought that every square is a rectangle and that not every rectangle is a square.

Analysis. As you can see from this episode, even though students are evidently quite familiar with the properties of the various Shape Makers, thinking about these classification problems is difficult for some of them. Although most of the students had correct answers, there was a variety of reasons for their answers, and there were several incorrect answers.

Most of the students decided that every square is a rectangle because (a) every shape made by the Rectangle Maker is a rectangle, and squares can be made by the Rectangle Maker; or (b) squares have all the properties of rectangles. (Similar reasoning was used to relate rectangles to parallelograms, and rhombuses to quadrilaterals.) Some students, however, argued that a square is not a rectangle because the Square Maker cannot make a traditional-looking rectangle. For them, this implied that squares cannot be made into rectangles, and therefore, squares can't be rectangles. Other students simply found it difficult to call a square a rectangle because *square* is the most familiar and precise term for the shape. They did not really consider the question as a logical classification problem, but rather thought of it as a naming issue.

The Mathematician's View

Mathematicians generally classify quadrilaterals hierarchically as shown here:

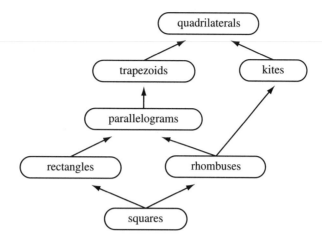

This diagram is interpreted as indicating, for example, that the set of all squares is a subset of the set of all rectangles; that is, every square is a rectangle. This is true because squares fit the definition of rectangles.

However, many students are quite uncomfortable with this classification system, and they often have convincing arguments to support their points of view. For instance, students frequently argue that calling a square a rectangle is not nearly as accurate—and probably would be counted wrong on a mathematics test.

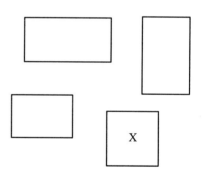

So we have to distinguish between the tasks of logical classification and maximally precise naming. These tasks have different purposes and contexts. Using logical classification to say, for example, that all squares are rectangles indicates a mathematical relationship between different classes of shapes. On the other hand, maximally precise naming is often required in communication. For instance, to indicate shape X at right, most people would refer to the "square."

The Special Case of Trapezoids. If you examine several geometry books, you will find two different definitions for *trapezoid*. One is the same as in this book—a quadrilateral with *at least* one pair of opposite sides parallel. The second defines a trapezoid as a quadrilateral with *exactly* one pair of opposite sides parallel. The former seems more elegant and is used in several popular high school geometry texts.

TRIANGLE
EXPLORATIONS

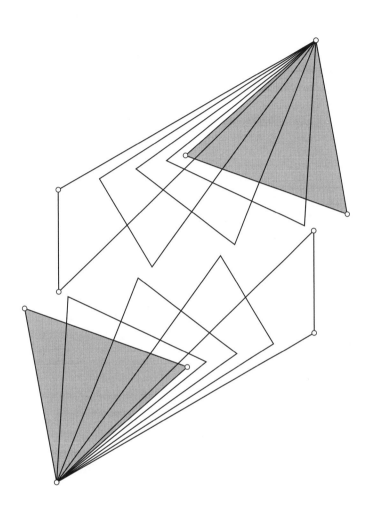

Making Pictures

Summary

Students figure out how right, equilateral, obtuse, isosceles, and general triangle Shape Makers can be used to make given pictures.

Mathematical Objectives

The goal of this exploration is for students to learn how the triangle Shape Makers operate and what kinds of shapes each makes. Students also begin to determine the properties of the different triangle Shape Makers.

Mac:

📁 Triangle Makers

    📁 Triangle Exploration 1

        ◆ Can You Make the Picture? #1

        ◆ Can You Make the Picture? #2

        ◆ Triangle Makers

Windows:

📁 TriMakrs

    📁 Tri_Exp1

        ◆ CYMP_1.gsp

        ◆ CYMP_2.gsp

        ◆ Triangle.gsp

Required Materials

Session	Student Sheet	SS#	Geometer's Sketchpad sketch
1 and 2	Can You Make the Picture? #1 & #2	43, 44	Can You Make the Picture? #1 & #2 Triangle Makers (Mac) CYMP_1.gsp and CYMP_2.gsp Triangle.gsp (Windows)

Use Recording Sheet for Can You Make the Picture? #1 and #2 on pages 222 and 223 as overhead charts for recording students' results. As shown here, you can record students' names and the letter of the shape they made with each triangle Shape Maker:

Student pair	Triangle Maker	Isosceles Triangle Maker	Equilateral Triangle Maker	Right Triangle Maker	Obtuse Triangle Maker
Jon & Jim	C	A	D	B	E

Activity: Can You Make the Picture?

Use Student Sheets 43 and 44, and sketches:

Mac: Windows:

📁 Triangle Exploration 1   📁 Tri_Exp1

    🔺 Can You Make the Picture? #1   🔺 CYMP_1.gsp

    🔺 Can You Make the Picture? #2   🔺 CMYP_2.gsp

Students Work in Pairs

➡ Distribute the Can You Make the Picture? #1 and Can You Make the Picture? #2 student sheets to the class.

When students open the sketches Can You Make the Picture? #1 and #2, the pictures on the student sheets appear on the screen. Students make the pictures on the screen using all five triangle Shape Makers: the Triangle Maker (for making any kind of triangle), the Isosceles Triangle Maker, the Equilateral Triangle Maker, the Right Triangle Maker, and the Obtuse Triangle Maker. (Acute and scalene triangles are dealt with later.)

➡ As students make the pictures, encourage them to think about and record properties of the triangle Shape Makers.

What do you notice about the Right Triangle Maker?

Why didn't you use the Equilateral Triangle Maker to make shape A?

Be sure to write down any interesting things you find out about the triangle Shape Makers. Be thinking about the properties of the triangle Shape Makers.

➡ Distribute a copy of the Conjectures and Queries sheet to each student. On it, students should record interesting ideas they discover about the various triangle Shape Makers. They should describe why they think these ideas are true or, if they are unsure about the validity of the ideas, why they are unsure.

➡ If students have extra time, you can have them make their own Can You Make the Picture? problems and trade with other student pairs.

To construct their own problems, students should use the five triangle Shape Makers in the sketch Triangle Makers to make a picture. They should then print two copies of their picture, one with the names of the triangle Shape Makers showing, one with the names hidden. To hide the names, students double click on the Hide names button. (If students have to transfer the sketch to another computer that is connected to a printer, they can choose **Save As** from the **File** menu. They'll have to give the sketch a unique name, perhaps using their own names or initials.) Students trade pictures, then use the sketch Triangle Makers to try to make the picture they have received. When they think they have the picture, they should check with the picture's creators to see if they used the Shape Makers in the same way.

Class Discussion

➡ After students have completed both student sheets, have a discussion about what they found.

Using the overhead charts described in the "Required Materials" section, have pairs of students tell which triangle Shape Makers they used for each shape. It is worthwhile to record all the *different* student answers, but not every pair's answers. Encourage students to object if they don't believe that a given Shape Maker will make a given shape, and to clearly explain their reasoning. Have the class attempt to resolve discrepancies.

➡ Ask students if they discovered any properties of the various triangle Shape Makers.

For instance, one student told his partner that the Isosceles Triangle Maker couldn't make shape B on Can You Make the Picture? #1 because shape B does not have a mirrored image and the Isosceles Triangle Maker always does. Another student, after several attempts to make shape A using the Equilateral Triangle Maker, said that it couldn't be done because the Equilateral Triangle Maker looked as if it had to have all of its sides equal.

➡ Ask students what similarities they see between the quadrilateral and triangle Shape Makers.

Responses can provide you with some initial indications of how students are thinking about Triangle Makers and Quadrilateral Makers. For instance, in one class, students said that the Isosceles Triangle Maker is like the Kite Maker because it mirrors itself, the Triangle Maker is like the Quadrilateral Maker because it makes all the different quadrilaterals, and the Equilateral Triangle Maker is like the Square Maker because it has equal sides.

2

Measured Triangle Makers

Summary

Students investigate the types of shapes that can be made by the measured Triangle Makers, which display the measures of angles and side lengths.

Mathematical Objectives

Students use the measured Triangle Makers and the concepts of angle and length measure to more precisely describe the operation of the triangle Shape Makers. They begin to formulate standard geometric properties of triangular shapes.

Mac:

- 📁 Triangle Makers
  - 📁 Triangle Exploration 2
    - 📄 M Equilateral Triangle Maker
    - 📄 M Isosceles Triangle Maker
    - 📄 M Obtuse Triangle Maker
    - 📄 M Right Triangle Maker
    - 📄 M Triangle Maker
- 📁 Demonstration Sketches
  - 📁 for Triangles
    - 📄 Folding Triangle Demo 1
    - 📄 Folding Triangle Demo 2

Windows:

- 📁 TriMakrs
  - 📁 Tri_Exp2
    - 📄 M_Eqtri.gsp
    - 📄 M_Isotri.gsp
    - 📄 M_Obtri.gsp
    - 📄 M_Rttri.gsp
    - 📄 M_Triang.gsp
- 📁 Demos
  - 📁 Triangle
    - 📄 FoldTri1
    - 📄 FoldTri2

Required Materials

Session	Student Sheet	SS#	Geometer's Sketchpad sketch
1 and 2	Investigating Measured Triangle Makers	45, 46	M Equilateral Triangle Maker, etc. (Mac) M_Eqtri.gsp, etc. (Windows)
3	The Sum of the Angles in a Triangle	47	M Triangle Maker (Mac) M_Triang.gsp (Windows)
4	How Are the Triangles the Same?	48	M Equilateral Triangle Maker, etc. (Mac) M_Eqtri.gsp, etc. (Windows)

Activity: Investigating Measured Triangle Makers

Use Student Sheets 45 and 46, and sketches:

Mac: Windows:

📁 Triangle Exploration 2   📁 Tri_Exp2

◈ M Equilateral Triangle Maker, etc.   ◈ M_Eqtri.gsp,etc.

Class Discussion

➡ Using a computer connected to an overhead display or gathering all students around a single computer, explain that there is a separate sketch for each measured triangle Shape Maker.

➡ Open the sketch M Triangle Maker. Have students explain the measurements displayed on the screen. As with the measured Quadrilateral Makers, these measurements indicate the lengths of the sides and the measures of the angles. Also have students discuss the symmetry buttons.

Students Work in Pairs

➡ Distribute the Investigating Measured Triangle Makers student sheets to the class.

Students predict which Triangle Makers will make triangles with certain properties, then check their predictions with the measured Triangle Maker sketches. See "Teaching Note: Investigating Measured Triangle Makers" on page 106 for an illustration of the types of ideas that occur to students in this exploration.

Class Discussion

➡ After students have completed their work, have them present their findings to the class.

For each problem on the student sheet, several student pairs should give and explain their answers, with other students challenging answers they disagree with. See "Teaching Note: Discussing Measured Triangle Makers" on page 106 for an illustration of how this class discussion might proceed.

➡ For each problem on the student sheets, ask students if they notice anything special about the triangles they made. Have students explain and discuss their ideas.

For example, as she was exploring problem 1, one student noticed that the Right Triangle Maker always has one side longer than the other two. Students have also noticed that triangles with three equal sides also have three equal angles, in fact, three 60° angles. In problem 2, students have noticed that triangles that have two 45° angles also have two equal sides and a right angle. In problem 3, students have seen that triangles with two equal angles have two equal sides. In problem 4, they have decided that triangles with two equal sides have two equal angles. If students haven't made such discoveries, however, don't worry about it. These ideas will be dealt with more explicitly in later sessions.

Also note that the discoveries that students make should be treated as *conjectures*, that is, as ideas that seem as if they might be true but have not been verified or proven. At this point, most of these discoveries are likely to have been derived empirically, probably using a limited number of examples. Students should not be convinced of their validity. Add the conjectures to the Conjectures and Queries sheet, discuss their conjectural nature with students, and encourage students to keep thinking about them.

Difficulties Due to the Precision of Displayed Measurements

Because the precision of displayed measurements in the *Shape Makers* microworld has been set to whole units, some apparently anomalous results might occur in this activity. For instance, some students conclude that the Right Triangle Maker can make triangles with all sides equal because they use the Right Triangle Maker to make a triangle with all three side lengths measuring 1 pixel. Such results should be explicitly discussed. By choosing **Preferences** from the **Display** menu and changing the precision of measurements from units to tenths (see "Mathematical Note: Precision of Measurements" on page 69 in Quadrilateral Exploration 4), students can see that the side lengths are not really all equal to 1 pixel. Another way to help students understand this particular result is to ask them whether they can make a right triangle with three 1-foot rulers.

Extension

This activity can easily be extended to include the concept of symmetry. For instance, you could add as a fifth problem: "(a) Predict which measured Triangle Makers can make a triangle with *at least one line of symmetry*. (b) Are there any triangle Shape Makers that always have three lines of symmetry?"

SESSION 3

Activity: The Sum of the Angles in a Triangle

Use Student Sheet 47 and sketch:

Mac:

Triangle Exploration 2

M Triangle Maker

Windows:

Tri_Exp2

M_Triang.gsp

As "Teaching Note: Discussing Measured Triangle Makers" on page 106 illustrates, many students will already have conjectured that the sum of the three angles in a triangle is 180°. The goal of this exploration is first to encourage all students to make this discovery, then to help students understand why it is true.

Students Work in Pairs

➡️ Distribute The Sum of the Angles in a Triangle student sheet to the class. Students use the sketch M Triangle Maker to investigate the sum of the angles of a triangle.

Keep in mind that students may get sums that do not equal 180° because the precision of angle measurements is still set to whole units. (Recall "Mathematical Note: Precision of Measurements" on page 69 in Quadrilateral Exploration 4.) By this time, students should know how to reset this precision to tenths. But suggest changing the precision only to those students who are having difficulty—that is, those who think that the sum is 180° but are puzzled because one or more of their sums do not equal 180°.

Class Discussion

➡ Have students share their ideas about the sum of the angles in a triangle. Most students will have concluded that the sum is 180°.

Some students, like Manuel in "Teaching Note: Discussing Measured Triangle Makers" on page 106, may justify their claim that the sum is 180° by saying that a triangle is half a quadrilateral, and because the sum of the angles in a quadrilateral is 360° you take half of that, or 180°, for a triangle.

➡ Using a computer connected to an overhead display or gathering all students around a single computer, use the sketches Folding Triangle Demo 1 and/or 2 to demonstrate how the three corners of a triangle can be folded to make a straight angle.

➡ Explain to students how the folds are made as shown below.

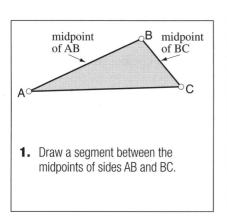

1. Draw a segment between the midpoints of sides AB and BC.

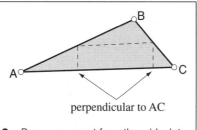

perpendicular to AC

2. Draw a segment from the midpoint of AB perpendicular to side AC. Draw a segment from the midpoint of BC perpendicular to side AC.

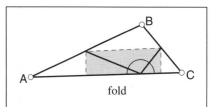

fold

3. Fold the three corners of the triangle along the segments you have drawn.

➡ Have students draw and cut out any triangle they wish. Then have them fold their triangles as described above.

➡ Have students explain the significance of their folding activity. Does this convince them that the sum of the angles in a triangle is 180°? Why?

Students might argue, for instance, that the three angles make a straight line. Because a straight line is half a full rotation, which is 360°, the line, and therefore the sum of the angles, is 180°.

Activity: How Are the Triangles the Same?

Use Student Sheet 48 and sketches:

Mac: Windows:

📁 Triangle Exploration 2   📁 Tri_Exp2

◈ M Equilateral Triangle Maker, etc.   ◈ M_Eqtri.gsp, etc.

Students Work in Pairs

➡ Distribute the How Are the Triangles the Same? student sheets to the class. Students use the measured Triangle Makers to discover the properties of isosceles, equilateral, right, and obtuse triangles.

That is, students attempt to determine what is the same about all shapes made with each type of triangle Shape Maker. The activity encourages them to think explicitly about the properties of *classes* of shapes, not particular shapes or Shape Makers.

Class Discussion

➡ Have students discuss each problem. As before, students should comment on, question, or support one another's answers. Students might name one or all of the properties listed here.

An *isosceles* triangle has at least two sides equal. (The angles opposite the equal sides are also equal.)

An *equilateral* triangle has all of its sides equal. (Because its angles are also all equal, and because the sum of the angles in any triangle is 180°, its three angles are each 60°.)

A *right* triangle has one right angle. The other two angles sum to 90°. The side opposite the right angle (the hypotenuse) is the longest side.

An *obtuse* triangle has one obtuse angle. The side opposite the obtuse angle is the longest side.

Assessment

Collecting the student sheets for this activity will give you a good idea about where students are in their thinking about classes of triangles. As you examine students' work, keep these questions in mind: What characteristics of triangles do students describe? Are these descriptions true or false? Are the characteristics that students explain genuine geometric properties that describe relationships between the parts of the shapes in terms of measures of parts? Do students name all of the properties for each triangle?

Investigating Measured Triangle Makers

A pair of students is working on the Investigating Measured Triangle Makers student sheets.

Problem 1: Predict which measured Triangle Makers can make a triangle with all of its sides equal.

Tom: Yes for Equilateral Triangle Maker because yesterday we found that it always has equal sides.

Dave: We think that the Isosceles Triangle Maker could probably make all of its sides equal because it is always supposed to have a mirrored image like the kite did. And we think that the side that's not in the mirrored image, we could probably get it to be the same length as the mirrored image sides.

Tom: I don't think the Right Triangle Maker can [pointing to the legs of triangle B on the Can You Make the Picture? #1 student sheet]. Because these two sides, they could be the same length. But when it is going diagonal to the corners [motioning along the hypotenuse of triangle B], the diagonal is longer.

Dave: For the Obtuse Triangle Maker I don't think it will work because it always has to have an obtuse angle, and an obtuse angle probably couldn't make all three sides equal. I'm not sure, but I just have a feeling that it won't.

Tom: I don't think it will either. It's sometimes sort of hard to get all of the sides the same length.

[The students now write reasons for their predictions.]

Tom: [For the Right Triangle Maker] No, because when there is a right angle in a triangle, there is always a diagonal side. The diagonal side is always longer than the two straight, nondiagonal sides.

Dave: For the Obtuse Triangle Maker we wrote no, because it makes odd shapes.

Tom: Well, the Obtuse Triangle Maker might be able to. I thought it couldn't. Because when there is an obtuse angle, it is sort of hard to get the sides the same length. I think you might be able to. [Note that Tom has changed his mind about the Obtuse Triangle Maker.]

Dave: [Dave has drawn an obtuse angle.] See, it is slanted outward. It is kind of hard to make a side the same length, going like that [drawing a third side to form an obtuse triangle], because it [the diagonal side] is sort of longer. It always has to be longer.

Tom: Yeah. The diagonal side.

Dave: So it is sort of like what we wrote for the Right Triangle Maker. That the slanted side is always longer.

Analysis. Dave and Tom have already discovered empirically that equilateral triangles have three equal sides. They see the Isosceles Triangle Maker as having a mirror image—that is, as being symmetric—and conclude that the "nonmirrored" side could be made the same length as the two "mirrored" sides. Their mental model for the Isosceles Triangle Maker seems quite accurate.

As the boys reflect on the Right Triangle Maker, they correctly conjecture that the side opposite the right angle—the hypotenuse—must be the longest side. They are later able to draw a similar conclusion for the Obtuse Triangle Maker, explicitly relating the two discoveries.

Problem 2: Predict which measured Triangle Makers can make a triangle with exactly two 45° angles.

Tom: Well, the Isosceles and the Equilateral Triangle Makers can, because if a triangle would have all of its side lengths equal, then all the angles are equal. And so you could probably change the angles when all the sides are equal.

Dave: Yeah, we noticed that all the angles were the same when you make all the sides equal.

Tom: If the side lengths were 60 pixels, the angles might be 45°. And if they are like 80 pixels, the angles might be 50°.

Dave: The Right Triangle Maker, I don't think it will. 'Cause with it always having a 90° angle, that interferes with it.

Tom: It might be able to. It always has to have a 90° angle. But the angles right here and right here, those might be the same [motioning to the two acute angles in a right triangle he has drawn].

Dave: Yeah, those might be the same. So I am guessing yes.

Tom: Right here, this would be another 90° angle [drawing a vertical line].

And the diagonal looks about like halfway over [drawing an arc from the vertical line to the hypotenuse]. Half of 90° is 45°.

Dave: I'm not sure about the Obtuse Triangle Maker.

Tom: [Tom draws an obtuse triangle on his paper.] It might. The angles right here and here might be the same [drawing little marks on the acute angles of the triangle].

Dave: I think it probably could. I'm not really sure though.

[Tom draws a horizontal line extending from one of the acute angles on his drawing of an obtuse triangle. He then draws an arc from the horizontal line to the side opposite the obtuse angle. See Illustration A below.]

Dave: We could do it [make two 45° angles] with the Obtuse Triangle Maker because the two angles look the same.

Tom: And again it looks about half [motioning to his picture, Illustration A]. Right here is the obtuse angle [pointing].

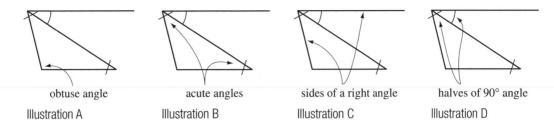

| obtuse angle | acute angles | sides of a right angle | halves of 90° angle |
| Illustration A | Illustration B | Illustration C | Illustration D |

And right here is another angle and another angle [motioning to the two acute angles]. [See Illustration B.]

And right here and here is a right angle [motioning to the left side of the triangle and the horizontal line]. [See Illustration C.]

And these look about half [motioning to the adjacent angles making up his 90° angle]. Like half of the 90° angle. [See Illustration D.]

[Dave and Tom start checking with the Triangle Makers.]

Dave: [Making different sizes of triangles with the Equilateral Triangle Maker] It's not changing the angles at all. You can't change the angles, just the pixels [continuing to manipulate the Equilateral Triangle Maker]. No, you can't. The angles don't change, only the length does.

Tom: Wow, that's surprising.

Dave: Yeah, I thought that it would.

Tom: Maybe because if the side lengths change and all of them are still the same, the angles don't change because it is just like the same shape only bigger.

[Dave now manipulates the Obtuse Triangle Maker, seemingly trying to make the obtuse angle smaller.]

Dave: It is not fair. No, it is not fair.

Tom: Try to move the other angle.

[Dave manipulates another control point.]

Dave: It is not fair, you can't go past 91.

[Dave continues to manipulate the Obtuse Triangle Maker.]

Tom: It won't work. All the angles have to add up to 180°. And if one is 91°, then the two other numbers in the ones places, the other angles, will have to add up to 9 because 9 plus 1 is 10 and that will get it to be 180°.

Analysis. When predicting which Triangle Makers can make a triangle with two 45° angles, the boys make some solid conjectures. Tom uses visual estimation to argue that the acute angles in a right triangle look like half of 90°, so must be 45°. However, he applies this same reasoning to mistakenly conclude that the Obtuse Triangle Maker can make two 45° angles. It is not until the boys manipulate the Obtuse Triangle Maker that they better understand the situation. At that time, they discover that one of the angles of the Obtuse Triangle Maker must always be more than or equal to 91°. From this and knowledge that the sum of the angles of a triangle must be 180°, they conclude that the two acute angles can't both be 45°.

We also see that some of the boys' intuitions lead to incorrect predictions. For instance, they think that a triangle that has all its sides the same length can have various angle measures, with the angle measures increasing as the side lengths increase. Again, their active manipulation with the Shape Makers helps them discover that this hypothesis is not viable. Furthermore, this manipulation prompts Tom to conjecture that angles will be the same size in triangles that are the same shape but not the same size, an experience that can support his future thinking about the concept of similarity.

In summary, we see that the combination of making predictions and checking these predictions with the Shape Makers enables students to develop, test, and refine their conceptions of the different types of triangles.

TEACHING NOTE
Discussing Measured Triangle Makers

Students are having a class discussion of their work on the Investigating Measured Triangle Makers student sheets.

Teacher: Laura, do you have one thing that you and Nanette found today that you thought was kind of neat?

Laura: If a triangle has two angles with the same degree, then it has to have, can have, two lines the same length. Because if you fold the shape in half, it has to be symmetrical.

Teacher: [Writing down Laura's idea on the board] Can or must?

Laura: Must.

Teacher: In question 2, the problem was to predict which measured Triangle Makers can make a triangle with two 45° angles. Let's see a show of hands for what you all said. [The teacher records the students' findings in a table.]

Triangle Maker	Yes	No
Isosceles	25	0
Equilateral	0	25
Right	21	4
Obtuse	0	25

Teacher: Alexis, how about you and Ada? Tell us what you said about the Right Triangle Maker and why you believe what you said.

Ada:	Well, we said yes. But also we found that when there were two 45° angles the other angle was always 90°. Because they always have to equal 180° when you add them.
Teacher:	What do the rest of you think about what Ada said?
Toshi:	We wrote that too. If you have two 45° angles, then automatically the other angle, the one that is not a 45° angle, is a 90° angle.
Manuel:	[Toshi's partner] It is called the Right Triangle Maker because it always has a right angle. And that is why it has to have a 90° angle in there.
Laura:	I have a question. I wondered why it has to be 180°.
Teacher:	Manuel gets to answer that first.
Manuel:	That's adding them up: 45 plus 45 plus 90 equals 180.
Laura:	I still don't see it.
Manuel:	If you have a right triangle like this [draws a square and one of its diagonals], then it would be half a square. And a square always has four 90° angles. And when you add those up, it's 360° and half of 360 is 180 and that's half a square.
Laura:	But why does it have to be 180°?
Manuel:	Because that shape is half of a square and half of 360 is 180.
Teacher:	[Drawing a square and one of its diagonals on the board] So you are saying to take this square and divide it into two right triangles. Here's one right triangle and here's the other right triangle. And this would be the hypotenuse.
Laura:	But then, does it always have to have 180° when you add it up? Does it have to be?
Teacher:	So your question is, Do right triangles always have to have 180°?
Laura:	No, do any triangles?
Teacher:	Do any triangles have to have 180°? I am going back to Alexis and Ada because they started this. Do you think any triangle has to have 180°, Alexis and Ada?
Alexis:	I think all of them.
Teacher:	Geoff, do you think all triangles have to have 180°?
Geoff:	No, I don't think that they all have to.
Teacher:	I think we need to stop for a second. How many of you think you have enough information to prove that they always have to be 180°? How many of you say no, I don't think I have enough yet? How many are not sure yet? Then I would like to stop our discussion of this idea for a bit so that you all have a chance to think about it and explore it on your computers.
	Okay, the four people who did not say that the Right Triangle Maker can make two 45° angles, can we hear from you so we can understand what happened?
Mark:	I don't really know, but we just tried for a while and we didn't get it.
Tanya:	It's the same here. We just kept trying and we couldn't get it.

Teacher: So do you want a little more time to work on that? Okay. So we will come back to that one again.

You all said that the Obtuse Triangle Maker couldn't make a triangle with two 45° angles. I want to know why you couldn't.

Kerry: Because it has to have over 90° in one angle. And 45 plus 45 is 90, so it would have to have under 45° angles.

Teacher: Could you say that again?

Kerry: The rest of the angles have to be under 45° because if the one angle was over 90°, then the other two couldn't both be 45°.

Teacher: So I hear you basing your argument on how many total degrees are in an obtuse triangle. What do you think?

Kerry: I think that there's 180°.

Brandy: I agree, and that's about what our logic was for that.

Teacher: So you think there are 180° in an obtuse triangle, just as other people think there are 180° in a right triangle?

Brandy: Yes.

Laura: We agree that all triangles have 180°, because we were experimenting with them and we used a calculator to find out.

Teacher: So you just wanted to support Manuel. Jackie, what do you and Tanya think about the Obtuse Triangle Maker? Why can't you get two 45° angles?

Jackie: Because 45 plus 45 equals 90. And the obtuse angle has to be greater than 90°.

Tanya: If you have two 45° angles, the other angle has to be 90° but it can't be 90°.

Teacher: Are you guys wondering about anything?

Laura: I'm just thinking about how we thought yes for our prediction, but when we tried it, it was 91° and 44° and 45° angles.

Dave: The obtuse can't have 45° angles because when you subtract 91 from the 180, you only have 89° left. So you can't get 45°. We tried it. We were surprised, because we wrote yes for our prediction. But we tried to pull it over and we could only get one 45° angle; we could only get the other one to 44°.

Teacher: What do you notice about all triangles that have two 45° angles, George?

George: We noticed that when two angles are 45°, then two side lengths are the same.

Teacher: George, why do you think that works?

George: Well, because like in the isosceles triangle, we said that if two lines are the same length, then the one line going across the bottom would make the two bottom angles the same because the lengths of the sides are the same.

[The teacher goes on to the answers for problem 3, "Predict which measured Triangle Makers can make a triangle with at least two equal angles." She again makes a table of student answers.]

Triangle Maker	Yes	No
Isosceles	25	0
Equilateral	25	0
Right	23	2
Obtuse	20	5

Teacher: Harrison, why do you think that everybody said yes for the Isosceles Triangle Maker?

Harrison: Because the Isosceles Triangle Maker is a lot like the Kite Maker in the fact that it mirrors itself. It always has a line of symmetry. Therefore, it's going to have two of everything. Two equal angles would be obvious if it has a symmetric image.

Jonathan: If you have two sides that are the same length and you just have one line connecting them, the two angles are going to be the same.

[Note that Harrison gives a reason for his assertion, whereas Jonathan merely restates it.]

Teacher: Do you agree, Tanya?

Tanya: Yeah!

Teacher: How would you say that?

Tanya: Two angles that are 45° each, then those two sides would have to be the same.

Kerry: I don't understand.

Jackie: If a shape has two equal angles, then the sides of the shape are equal too.

Kerry: Okay.

Teacher: Actually, that is what Tanya said, but Jonathan said the reverse, didn't he? Jonathan said if the two side lengths are the same, the two angles have to be the same. And Tanya and Jackie said if the two angles are the same, the side lengths have to be the same. If it follows one way, does it have to follow the other way?

Kerry: I don't think it always does, but I think in this case it does.

Jackie: I'm starting to have second thoughts about what Jonathan was saying. He said that when two sides are the same, the angles would be the same. What about the Right Triangle Maker?

Jonathan: Here's a right angle, and these two sides are the same length. [He uses one hand to show a right angle, then makes a diagonal with his other hand.]

If this side is the same as this side [making the right angle again with his left hand, he then points to the two equal sides coming off of the right angle], then I think this angle and this angle are going to be the same [motioning to the two nonright angles].

Teacher: [Referring to a picture of a right triangle on the board] So you are saying if this would have been the same length and this would have been the same, then those two angles would be the same. And you are not sure that is true, Jackie?

Jackie: I'm not sure.

Teacher: Have you written that down? You might want to check it out. In fact, you've come up with lots of great conjectures. But I think that many of you are not convinced about some of these ideas. We'll write these ideas on our Conjectures and Queries sheet. When you have time, see if you can convince yourself that these conjectures are true or false. You can use the Shape Makers to help you. But you should also try to use logic and things you already know.

Analysis. The students have generated many interesting conjectures. The teacher acts as a facilitator, asking for ideas, tabulating results, and requesting input about conjectures—including who agrees and disagrees and why ideas might be true or false. She encourages the students to discriminate between what they know and what they suspect (especially concerning the issue of the sum of the angles in a triangle, which will be dealt with explicitly in the next session). The teacher subtly guides the focus of students' attention and the direction of student discourse.

Justifications of statements vary considerably among students. Harrison's argument for why an Isosceles Triangle Maker can make two equal angles is quite sound. Jonathan's justification of this idea, on the other hand, is merely a restatement of the conjecture—he offers no real proof that his statement is true. Manuel's argument about why the sum of the angles of a triangle is 180° is sound, but it is incomplete. It applies only to right isosceles triangles, not to right triangles in general or to all triangles. Finally, the student arguments about why an Obtuse Triangle Maker can't have two 45° angles are correct but, for the most part, not clearly explained.

Other Kinds of Triangles

Summary

Students determine what kinds of triangles cannot be made with the Obtuse, Right, Equilateral, and Isosceles Triangle Makers. They are then introduced to two additional types of triangles, scalene and acute.

Mathematical Objectives

Students analyze the properties of the Obtuse, Right, Equilateral, and Isosceles Triangle Makers in a way that leads them to see that there are two additional types or classes of triangles. They define these new types by their properties, then analyze certain interrelationships between the classes of triangles. Students are thus able to completely classify triangles by side length and angle measure.

Mac:

Triangle Makers

Triangle Exploration 3

Other Kinds of Triangles 1

Other Kinds of Triangles 2

Windows:

TriMakrs

Tri_Exp3

Ot_Tri1.gsp

Ot_Tri2.gsp

Required Materials

Session	Student Sheet	SS#	Geometer's Sketchpad sketch
1	Other Kinds of Triangles	49–54	Other Kinds of Triangles 1 & 2 (Mac) Ot_Tri1.gsp and Ot_Tri2.gsp (Windows)

Activity: Other Kinds of Triangles

Use Student Sheets 49–54, and sketches:

Mac: Windows:

📁 Triangle Exploration 3   📁 Tri_Exp3

    ◆ Other Kinds of Triangles 1   ◆ Ot_Tri1.gsp

    ◆ Other Kinds of Triangles 2   ◆ Ot_Tri2.gsp

Students Work in Pairs

➡ Have students use the Triangle Makers to answer the questions posed on the Other Kinds of Triangles student sheets. They are to make conjectures about—then investigate with measured triangle Shape Makers—what types of triangles can't be made by either the Equilateral or Isosceles Triangle Maker, then by either the Obtuse or Right Triangle Maker.

Class Discussion

➡ Have students discuss their findings. As they do, introduce the two new types of triangles, scalene and acute.

For problem 1, students should conclude that triangles that cannot be made by either the Equilateral or Isosceles Triangle Maker must have no sides congruent. Be sure all students understand why. Once students have drawn this conclusion, tell them that such triangles are called *scalene triangles*.

For problem 2, students should conclude that triangles that cannot be made by either the Obtuse or Right Triangle Maker must have all acute angles. Be sure all students understand why. Once students have drawn this conclusion, tell them that such triangles are called *acute triangles*.

➡ Have students clarify the idea of classifying triangles by their angles.

It is important that students see that a triangle is a right triangle if it has *one* right angle; it is obtuse if it has *one* obtuse angle; but it is acute only if *all* of its angles are acute. You might help students clarify this idea by asking them:

> *How do you know if a triangle is a right triangle? How many angles have to be right angles?*
>
> *How do you know if a triangle is an obtuse triangle? How many angles have to be obtuse?*
>
> *How do you know if a triangle is an acute triangle? How many angles have to be acute?*
>
> *Why does a right triangle have to have just one right angle and an obtuse triangle just one obtuse angle, while an acute triangle has to have three acute angles?*

See "Teaching Note: It Takes More Than a Pretty Smile to Be an Acute Triangle" on page 113 for an illustration of how students deal with this issue.

➡ Have students discuss the issue of combined angle/side length classifications. In problems 3 and 4, students are asked to think about classifying triangles by length of sides and measure of angles. Some classification combinations are impossible. For instance, there are no obtuse equilateral triangles, because all the angles in an equilateral triangle are acute.

Angle/Side Length Triangle Classifications

You might want to discuss with students the standard way that triangles are classified. Every triangle can be classified by angle measure and by side length. There are nine possible ways to combine these classifications. As problem 4 illustrates, however, some of these combinations are in fact impossible. As an exercise, you can have students write all nine possible combinations, then identify those that actually occur.

TEACHING NOTE

It Takes More Than a Pretty Smile to Be (an) Acute Triangle

These students have explored the five original measured Triangle Makers. They are now discussing the idea of a triangle that has all acute angles.

Teacher: What do you think you call a triangle that has three angles that are less than 90°?

Arnie: An acute triangle.

Teacher: Why would you guess that?

Arnie: Because an acute angle, it's an angle with less than 90°.

Alexis: But you can't have a triangle with three acute angles.

Li Chen: An equilateral always has three acute angles.

Alexis: Oh!

Teacher: Why were you thinking you couldn't?

Alexis: Because I thought it was really acute angles, real small.

Teacher: So now we have a new kind of triangle, an acute triangle. If you were going to list a property of an acute triangle, what do you think you would list?

Alexis: It has three acute angles.

George: It could be like an obtuse. And it might just have two acute and one obtuse.

Teacher: What about that? Do you think an acute triangle could have one obtuse and two acute angles?

Nanette: That's the same as the obtuse triangle.

Teacher: George, how would we differentiate between an obtuse triangle and an acute triangle?

George: Three acute angles.

Teacher:	This is a strange one, isn't it? So, George, would you agree that it always has to have three acute angles? Because as soon as you get two acute angles and one obtuse angle, what do you end up with?
Nanette:	An obtuse triangle.
Teacher:	Does anybody disagree?
Dave:	Every triangle has an acute angle. So you could call the Triangle Maker the Acute Triangle Maker, because any triangle has an acute angle.
Brandy:	I don't think we could, because we said an acute triangle has to have all three of its angles less than 90°. And the Triangle Maker can make any triangle.
Teacher:	What I'd like you to do is take about 30 seconds to talk it over with your partner. Does it make sense to call the Triangle Maker an Acute Triangle Maker?

[After students have discussed their ideas with their partners, the class discussion resumes.]

Tom:	No, because the Triangle Maker can make any triangle, and an Acute Triangle Maker could only make triangles with three acute angles. An Acute Triangle Maker couldn't make obtuse triangles.
Toshi:	I agree. And it could not make right triangles.
Harrison:	The Triangle Maker doesn't always have all acute angles. The Acute Triangle Maker does.
Jon:	I say no, because the Triangle Maker can make triangles without three acute angles.
Teacher:	What do you think the rule or characteristic of acute triangles is? To be an acute triangle, what must it have?
Alexis:	Three acute angles.
Dave:	But how come the Obtuse Triangle Maker only has to have one obtuse angle? I thought we could call the Triangle Maker the Acute Triangle Maker because I was thinking that every triangle has at least one acute angle.
Teacher:	Good question. Why does the Obtuse Triangle Maker have just one obtuse angle?
Dave:	I know why! Because an obtuse angle points outward, and you can't make a triangle with three obtuse angles; the sides wouldn't come together.
Teacher:	And why does an acute triangle have to have three acute angles, why can't it be just one or two?
Alexis:	Because a right triangle and an obtuse triangle both have two acute angles. So the only way to get something different is to have all three angles acute.

Analysis. These students are struggling with the fact that an acute triangle must have all three angles acute, whereas an obtuse triangle and a right triangle must have only one obtuse or one right angle, respectively. The episode clearly shows that something we, as knowledgeable adults, take as natural can seem quite strange and difficult for students to make sense of and accept. It is only through class discussions like the one above that students have the opportunity to sort out and make sense of such ideas. These students needed to reflect hard on the logic of the definitions for acute, obtuse, and right triangles.

Making Shapes from Triangle Pairs

Summary

Students make quadrilaterals by rotating or reflecting triangles about their sides. They then identify the types of quadrilaterals made and justify their identification claims.

Mathematical Objectives

This exploration encourages students to use analytic, property-based thinking about both triangles and quadrilaterals. It encourages students to integrate the van Hiele level 2 thinking they have developed for quadrilaterals and triangles.

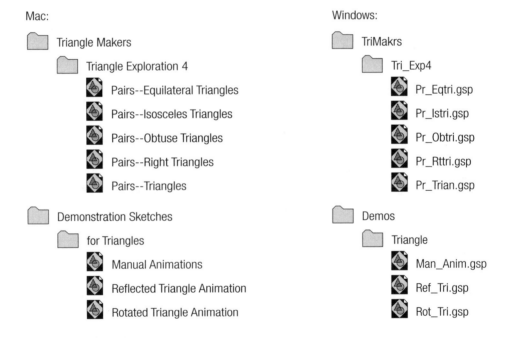

Mac:

📁 Triangle Makers

   📁 Triangle Exploration 4

      ◆ Pairs--Equilateral Triangles

      ◆ Pairs--Isosceles Triangles

      ◆ Pairs--Obtuse Triangles

      ◆ Pairs--Right Triangles

      ◆ Pairs--Triangles

📁 Demonstration Sketches

   📁 for Triangles

      ◆ Manual Animations

      ◆ Reflected Triangle Animation

      ◆ Rotated Triangle Animation

Windows:

📁 TriMakrs

   📁 Tri_Exp4

      ◆ Pr_Eqtri.gsp

      ◆ Pr_Istri.gsp

      ◆ Pr_Obtri.gsp

      ◆ Pr_Rttri.gsp

      ◆ Pr_Trian.gsp

📁 Demos

   📁 Triangle

      ◆ Man_Anim.gsp

      ◆ Ref_Tri.gsp

      ◆ Rot_Tri.gsp

Required Materials

Session	Student Sheet	SS#	Geometer's Sketchpad sketch
1	Making Shapes from Triangle Pairs	55–57	Demonstration sketches Pairs--Triangles (Mac) Pr_Trian.gsp (Windows)
2–4	Making Special Quadrilaterals from Triangle Pairs	58–62	All Pairs--Triangles, etc. (Mac) Pr_Rttri.gsp, etc. (Windows)
5	What Kind of Triangle? What Kind of Quadrilateral	63, 64	

Introduction to Making Shapes from Triangle Pairs

Class Discussion

➡ As described below, demonstrate—first with a computer, then with paper folding—how triangles can be reflected and rotated about their sides to form quadrilaterals or triangles. Then show students how the sketch Pairs--Triangles is used.

Computer Demonstration

➡ Using a computer connected to an overhead display or gathering all students around a single computer, use the sketches Rotated Triangle Animation, Reflected Triangle Animation, and Manual Animations to illustrate how a triangle can be rotated and reflected about one of its sides to form a figure whose outline is a quadrilateral or a triangle.

➡ **Reflection.** Double click on the Reflect about CD button in the sketch Reflected Triangle Animation to illustrate how a triangle can be reflected about one of its sides to form a quadrilateral.

In this demonstration, we start with triangle ABC. When this triangle is reflected about side BC and the two triangles are combined, the result is the figure pictured in Figure TE 4.1. The outline of this combined figure, as shown by the dark outline in Figure TE 4.2, is the focus of our discussion. This outline is a quadrilateral.

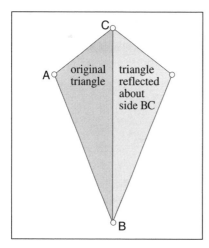

Figure TE 4.1   Figure TE 4.2

➡ **Rotation.** Double click on the Rotate about BC button in the sketch Rotated Triangle Animation to illustrate how a triangle can be rotated about one of its sides to form a quadrilateral.

In this demonstration, the triangle is rotated about side BC. (Actually, it is rotated 180° about the midpoint of side BC. But students really don't need to know this.)

When the original and rotated triangles are combined, the result is the figure pictured in Figure TE 4.3. The outline of this combined figure, as shown by the dark line in Figure TE 4.4, is the focus of our discussion. This outline is a quadrilateral.

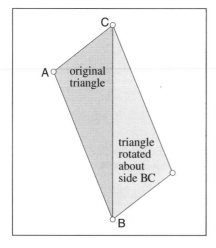

Figure TE 4.3

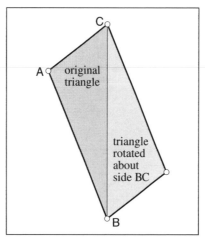

Figure TE 4.4

The reflection and rotation motions can also be illustrated with the sketch Manual Animations shown in Figure TE 4.5. To show the reflection of triangle ABC about side BC, drag the control point from left to right and back along the dotted "reflection" segment. To show the rotation of triangle ABC about side BC, drag the control point along the "rotation" arc. You can change the shape of triangle ABC by dragging its control points, but doing so causes the reflection control segment to move, so you have to be careful not to move it off the screen.

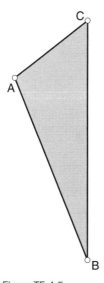

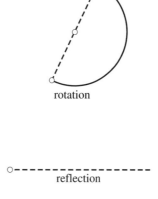

Figure TE 4.5

Paper and Pencil Demonstration

➡ Illustrate these reflections and rotations using paper cutting.

To show a reflection:

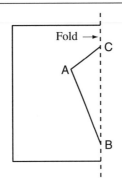

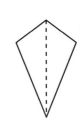

1. Fold a sheet of paper in half. Draw a triangle with side BC along the fold line. While the paper is still folded, cut out the triangle.

2. Open up the fold to see the figure made by the triangle and its reflected image.

To show a rotation:

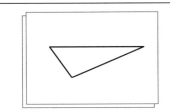

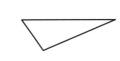

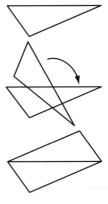

1. Position two sheets of paper so that one is on top of the other. Draw a triangle on the top sheet.

2. Cut out the triangles simultaneously, while the two sheets are still together. (You might label the vertices of each triangle A, B, and C to keep track of them during the rotation.)

3. Rotate the top triangle 180°, until one of its sides overlaps the corresponding side on the bottom triangle.

Demonstrate the Sketch Pairs--Triangles

➡ Illustrate to students how the sketch Pairs--Triangles is used.

As shown in Figure TE 4.6, the sketch Pairs--Triangles contains a measured Triangle Maker. Students move the control points on the yellow Triangle Maker to change its shape. By double clicking on the Show reflection in AC button, students can see the triangle reflected about side AC. Measurements for the newly created sides and angles are also displayed. (There are also buttons for reflecting about the other two sides of the triangle and for rotating the triangle about its three sides.)

Note that on the computer the image of a point under a reflection or rotation is labeled with a prime ('). For instance, the image of vertex A under a reflection is labeled A' (said "A prime"). Be warned, however, that the point A' that results from a reflection of point A and the point A' that results from a rotation of point A will not, in general, be the same.

Length(AC) = 100 pixels	Angle(BAC) = 41°
Length(CB) = 77 pixels	Angle(ABC) = 58°
Length(BA) = 117 pixels	Angle(BCA) = 82°

Length(CB') = 77 pixels
Length(AB') = 117 pixels
Angle(AB'C) = 58°
Angle(B'AC) = 41°
Angle(ACB') = 82°

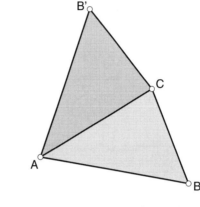

▲ Show reflection in AC
△ Hide reflection in AC

Figure TE 4.6

Note that angles are now named with three letters instead of one. This is necessary because, for instance, at point C there are three angles: BCB', ACB, and ACB'. So saying "angle C" would be ambiguous. In this naming scheme, the point at the vertex of the angle must be named by the second letter, but which letter occurs first or last doesn't matter. Ask students what *vertex* means to ensure that they recall the meaning of this term.

➡ Discuss this new angle-naming scheme with students by asking questions:

What's different about the angle measurements on the screen?

Why do you think three letters are used instead of one?

For which vertices is this important?

Which letter occurs in the middle?

Show me angle ACB. Show me angle BCB'.

[Pointing to angle ACB'] What would we call this angle?

➡ Hide the reflection and change the shape of triangle ABC. To help students better understand the effects of reflecting and rotating a triangle about its sides, ask them to predict the effects of various reflections and rotations. Check their predictions by showing them what actually happens on the computer. For instance, referring to the sketch in Figure TE 4.7, you might ask students:

Where do you think the reflection of angle B will appear when triangle ABC is reflected about side AC?

Where do you think the reflection of angle B will appear when triangle ABC is reflected about side AB?

Can you draw what the screen will look like after triangle ABC is reflected about side AC?

Can you draw what the screen will look like after triangle ABC is rotated about side AB?

Repeat for several other triangle shapes.

Length(AB) = 81 pixels	Angle(BCA) = 100°
Length(AC) = 117 pixels	Angle(BAC) = 37°
Length(BC) = 72 pixels	Angle(ABC) = 43°

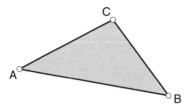

▲ Show reflection in AC	▲ Show rotation in AC
△ Hide reflection in AC	△ Hide rotation in AC
▲ Show reflection in AB	▲ Show rotation in AB
△ Hide reflection in AB	△ Hide rotation in AB
▲ Show reflection in BC	▲ Show rotation in BC
△ Hide reflection in BC	△ Hide rotation in BC

Figure TE 4.7

Activity: **Making Shapes from Triangle Pairs**

Use Student Sheets 55–57, and sketches:

Mac: Windows:

📁 Triangle Exploration 4   📁 Tri_Exp4

◈ Pairs--Triangles   ◈ Pr_Trian.gsp

Students Work in Pairs

➡ Distribute the Making Shapes from Triangle Pairs student sheets to the class. Students use the sketch Pairs--Triangles to explore the problem "What kinds of shapes do you get when you reflect or rotate a triangle about one of its sides and combine the two triangles?"

➡ Have students make each different type of shape they find on the computer by cutting triangles out of paper.

Although some students might want to use a ruler to measure the shapes on the screen, it is not necessary that they duplicate these shapes exactly. They only need to cut out a shape of the same type; usually, drawing a shape that looks close to the one they made on the screen will do. Students tape their cutout triangles to their student sheets, then describe the shapes and the motion (reflection or rotation) they used to make them. For example, they might say that they formed a shape by reflecting the triangle about one of its sides.

Class Discussion

➡️ Have students describe their findings.

➡️ For a couple of the outline shapes that students have made, have them describe how they know that the shape is what they say it is.

For example, if they say that the shape formed by the two triangles is a parallelogram, they should describe how they would "prove," or convince others, that it is a parallelogram. They might argue that the opposite sides are equal and adjacent angles add to 180°, two properties that they might have found when investigating quadrilaterals. However, there is no need to force students to be this analytic at this time; they will have further opportunities for this type of analysis in the next several sessions.

SESSIONS 2–4

Activity: **Making Special Quadrilaterals from Triangle Pairs**

Use Student Sheets 58–62, and sketches:

Mac: Windows:

📁 Triangle Exploration 4   📁 Tri_Exp4

◈ Pairs--Right Triangles, etc.   ◈ Pr_Rttri.gsp, etc.

Students Work in Pairs

➡️ Distribute the Making Special Quadrilaterals from Triangle Pairs student sheets to the class.

➡️ Have students use the sketches Pairs--Equilateral triangles, Pairs--Isosceles triangles, Pairs--Right triangles, and Pairs--Obtuse triangles to decide which types of special triangles can be used to make special kinds of quadrilaterals, such as rectangles and parallelograms.

As in the last activity, students drag the control points on a Triangle Maker to change its shape, then double click on the Show and Hide buttons to see the quadrilaterals formed by reflecting or rotating the triangles.

Students first *predict* (using drawing, paper cutting, or visualization) whether a given quadrilateral can be made by the particular Triangle Maker. They then *check* with the Triangle Makers in the Pairs sketches. For instance, students first predict whether the Right Triangle Maker can be used to make a square, then check their prediction with the sketch Pairs--Right triangles sketch. They then predict whether the Right Triangle Maker can be used to make a rectangle that is not a square, then check their prediction on the computer, and so on.

For predictions and checks, students circle Y if the special quadrilateral can be made, N if it can't. If their check shows that a special quadrilateral can be made, students are to tell *how* they made it by circling "rotate" or "reflect" and the side they reflected or rotated about. Finally, after checking with the Shape Makers, students are to explain:

(a) if they made the quadrilateral, why they think it is what they say it is (for example, if they think they have formed a rectangle, they must convince somebody that the shape is a rectangle); (b) if they couldn't make the quadrilateral, why they couldn't make it. In either case, if students don't refer to measurements and properties, encourage them to do so.

> *How can you prove that this shape is a parallelogram?*
> *Why don't you think the Obtuse Triangle Maker can be used to make a rectangle?*

➡ Encourage students to notice conditions that must be placed on various Triangle Makers to make special quadrilaterals. For instance, if students find that a Right Triangle Maker flipped about its hypotenuse makes a square, encourage them to think about what kind of right triangle was used. In fact, it is a right triangle with its shorter sides equal—that is, an *isosceles* right triangle flipped about its hypotenuse makes a square.

➡ To help clarify this task for students, you might want to have a class discussion after all students have finished the first problem, the Right Triangle Maker. They will then have a better idea of what is expected of them in subsequent problems.

Student Strategies for Thinking About the Result of Reflecting/Rotating a Triangle

Because it is difficult for students to draw and, especially, to visualize the result of reflecting or rotating a triangle, some students think about the result by first drawing a quadrilateral, then drawing one of its diagonals. They then inspect the resulting triangles to see what type they are. For instance, a student trying to decide whether an Isosceles Triangle Maker can be used to make a rectangle might draw a rectangle and a diagonal.

The student would then examine, say, the bottom triangle and try to decide whether it is isosceles, either visually or using a ruler. Although we may think that students should see the same things we see in such diagrams, keep in mind that what we see is determined to a great extent by what we think is true. So, for instance, while we know that these triangles are not isosceles, some students will conclude that they are.

As students gain more experience with this activity, many will start making impressive deductions about the possibilities. For instance, one pair of students reasoned that in order for the Isosceles Triangle Maker to make a rectangle, it would have to have a right angle. This would mean that the other two angles (the base angles) would have to be 45°. But if these two angles are congruent, then the sides opposite them would be congruent. So the only way that the Isosceles Triangle Maker could be used to make a rectangle would be for it to make a square. Since they were trying to find a rectangle that wasn't a square, they concluded that it couldn't be done.

Unsystematic Approaches

Also keep in mind that many students will not approach this activity completely systematically. You should remind students to search all the possibilities—reflections and rotations about all three sides of the triangle. Don't, however, make them systematically fill out a table of all possibilities or demonstrate going through all the possibilities. Students need to develop their own methods of solution, not blindly follow a system that we give them but that makes no sense to them. Student searches will gradually become more systematic as students come to see more and more possibilities and as they recognize, independently or in class discussions, the deficiencies in unsystematic approaches.

Class Discussion

➡ For each Triangle Maker and each special quadrilateral on the student sheets, have several students explain their findings.

If students made the given quadrilateral, they should first explain how they made it, then attempt to convince the class that it is what they say it is. They should also explain whether there is anything special about the shape of the Triangle Maker needed to make the quadrilateral (e.g., only an isosceles Right Triangle Maker can be used to make a square). If students couldn't make the quadrilateral, they should explain why they couldn't make it.

Note that students will have to combine the measures of some of the two triangles' angles to get the angles of the resulting quadrilateral. Have them describe how they did this.

For an illustration of how a class discussion on making special quadrilaterals might go, see "Teaching Note: Making a Rectangle from Two Right Triangles" on page 125. For a description of how students might justify their claims, see "Teaching Note: Proving It's a Square; Proving It's a Parallelogram" on page 128.

At times, you will find it useful for you or your students to illustrate what is being talked about, using actual examples from the Pairs sketches. You can do this by (a) using a computer connected to an overhead display or gathering all students around a single computer and showing the sketch, or (b) printing a sketch and making either an overhead transparency or individual copies for students.

SESSION 5

Activity: What Kind of Triangle? What Kind of Quadrilateral? (Assessment)

Use Student Sheets 63 and 64.

Students Work Alone

➡ Have each student complete the What Kind of Triangle? What Kind of Quadrilateral? student sheets individually, without the use of computers. After you have collected students' sheets, you might have a class discussion on this problem.

Assessment

 Analyze students' work on the student sheets.

Guidelines for Analysis of Students' Work

Students should justify that triangle ABC is isosceles and acute by correctly referring to the appropriate measurements: Side AB has the same length as side AC—118 pixels; angle ABC and angle BCA both measure 71°, and angle BAC measures 38°.

Students should justify that triangle ABC was rotated about side AC by noting that it couldn't have been reflected, because then angles BAC and B'AC would be congruent, which they are not.

Students should justify that quadrilateral ABCB' is a parallelogram by correctly referring to appropriate measurements and properties of parallelograms. For example, they might use the measurements to show that opposite sides of the quadrilateral are congruent and that opposite sides are parallel. The latter can be shown by noting that alternate interior angles are congruent; for example, angle BAC is congruent to angle B'AC.

Some students may incorrectly identify the shapes because they have incomplete knowledge of their properties. A few students may say that the shapes *look like* an acute isosceles triangle and a parallelogram because they have not completed the transition to van Hiele level 2 thinking.

Most students will try to prove that quadrilateral ABCB' has *all* the properties that they have agreed upon for parallelograms—for example, opposite sides congruent and parallel, opposite angles congruent—and similarly for acute and isosceles triangles. Not only are these students showing that they understand what the properties of these shapes are, they are exhibiting evidence of being at van Hiele level 2.

You may also find, however, that some students are beginning to focus on *sufficient* conditions for identifying shapes. For instance, they might say that all you need to prove for a shape to be a parallelogram is that the opposite sides are congruent, because this implies that the opposite sides are parallel. You might ask them why they think this is true. In response, students typically give a visual description of the possibilities of movement for the sides of a quadrilateral. They reason using the mental models they have formed from their work with the various Shape Makers. These students are showing definite signs of van Hiele level 3 thinking.

See "Teaching Note: Proving It's a Square; Proving It's a Parallelogram" on page 128 for more discussion of the ways students might think about these problems. The section "Proving It's a Parallelogram" in this teaching note also has suggestions about how to help students moving into level 3 refine their thinking.

Class Discussion

If you have time, it is productive to have students discuss their work on this assessment sheet in a whole-class discussion. Encourage students to try to reach a consensus on the acceptability of various "proofs" for the different problems. Some fascinating discussions can result:

Making a Rectangle from Two Right Triangles

Students are having a class discussion on the Making Special Quadrilaterals from Triangle Pairs student sheets.

Teacher: I'd like to have a little discussion about whether the Right Triangle Maker can make a rectangle that is not a square; 22 people said yes, and 4 people said no. Can I hear from one of these 4 people?

George: We could get a parallelogram, but we couldn't make 90° angles.

Tina: We couldn't get 90° angles.

Teacher: Any thoughts from the 22 yes-vote people?

Lauren: How could you get the square?

George: We made 90, 45, and 45 in our triangle, and it made a square.

Alex: At first, Steve and I were thinking to make 90, 45, and 45, but actually that is not what you want, because then you get exactly a square.

Mark: On ours it was 70 and 20 [the angles].

Teacher: So you made a right triangle that didn't have equal angles, and that worked. And did you do a rotation or a reflection?

Alex: We did a reflection.

Mark: A rotation.

Teacher: If you reflected, where would your 70s go?

Students: Together.

Teacher: Together, wouldn't they? So would that work?

Students: No.

Teacher: What do you think, Kerry?

Kerry: It would have to be a rotation, because then the 70s would go with your 20s.

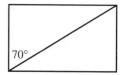

Teacher: So you're saying that if this is a 70 [labeling one angle], if I rotate it, where would my other 70° angle be?

Students: The other end.

Teacher: It is going to be up here, isn't it? And what is this angle [pointing]?

And what is this angle?

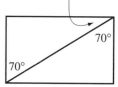

Students: 20°.

Teacher: So you want to rotate the 70° angle so that it is adjacent to the 20° angle. And if you did a reflection, what angles would be adjacent?

Students: 70 and 70.

Teacher: 70 and 70. So then what would you get?

Students: 140.

Teacher: 140, so would you get a rectangle?

Students: No.

Teacher: Good. It's important to think of where your angles end up on your reflections versus rotations.

Harrison, I also want to pick up on your idea that if you have 45° angles, you always get a square.

Harrison: Because two 45s is 90. And if you put the two 45° angles together, it would be a 90° angle. A square has four 90° angles.

Teacher: Why do I always get a square?

Harrison: The two sides have to be equal for the two 45° angles to be able to ever meet each other and make a whole solid shape.

Teacher: That's what I thought I heard you talking about yesterday. Harrison was maintaining that if you have two 45° angles, then you automatically have two sides that have equal lengths, and that is why you get the square every time. That's an interesting thought for you to check out to see if it is true.

Harrison: I have another theory. Any reflection that is not another triangle, you know we talked about how if you reflect a right triangle on AC, you just get another triangle because of the two right angles. I think any reflection, unless it is a triangle, is a kite.

Teacher: Do you mean any reflection of a Right Triangle Maker, or any triangle Shape Maker?

Harrison: Well, I was thinking about right triangles. But I think it is true about all triangle Shape Makers.

Teacher: That's interesting, Harrison. What do you all think about this? We should write this idea down on our Conjectures and Queries sheet and check it out when we have time. In fact, I think we need to reflect a bit more on several of the ideas we have talked about today. So what you are going to do is go back to your computers and use the Right Triangle Maker to double-check your answers for rectangle, parallelogram, kite, and rhombus. If you have extra time, you can check out some of the other ideas that have been mentioned, such as Harrison's.

[The teacher has different student pairs investigate making these four shapes with the Right Triangle Maker. Dave and Tom attempt to make a rectangle.]

Tom: Wouldn't that be a rectangle? [No rotation has been shown yet; the boys are looking only at the triangle.]

Dave: Well, let me just pull this [control point] out some more.

Tom: Yeah, it would be a rectangle.

Dave: [Showing the rotation of the triangle] Yeah, it's a rectangle.

Tom: [Pointing to the rectangle on the screen] It's a rectangle [pause; then, pointing more specifically to the angles in the rectangle and triangles], because 67 and 23 equal 90. Yeah, it's a rectangle.

Dave: We've got a rectangle because here is the 90° angle right here [pointing to the bottom right angle in the rectangle]. If this angle right here is 45° and this angle is 45° [motioning to the two nonright angles in the bottom right triangle], then it would be a square, if you did the rotation of it. But if it is anything over 45, it could be 46 and 44, and then you do the rotation on it, it would still be a rectangle. So we did it with 67 and 23, and it is a rectangle.

[The teacher asks the groups who worked on getting a rectangle from the Right Triangle Maker how they got one. All three groups did it by a rotation. Alex questions the shape on Manuel and Toshi's computer.]

Alex: That looks like a square.

[Toshi and Manuel tell the measurements of the sides of the square, showing that they are not all equal.]

Teacher: You know what, Alex, I'm glad you said that. How do we know? What is the real proof that it is a square or a rectangle?

Alex: The side lengths.

Teacher: What are the real lengths?

Manuel: 124 and 122.

Teacher: So the horizontal is longer, even though all sides look the same. Alex brought up a very important point. Can you just rely on visual clues for determining shapes? As Alex looked over at that screen, to him it appeared as if it were a square. But when he got up close and checked out the facts, he found out it was a rectangle. So we really need to look carefully at the data, don't we?

            There were eight people who said that you could not make a parallelogram with the Right Triangle Maker.

Alexis: We changed our mind.

Teacher: What made you change your mind?

Alexis: Alex and Steve showed us. See, before, we didn't think to reflect about a line that wasn't a hypotenuse.

Teacher: So you think that to get a parallelogram you have to reflect it about a nonhypotenuse side.

Kelly: Do you mean reflection or rotation?

Alexis: Rotation. That's what I meant.

Analysis. In the first part of this class discussion, the teacher helped focus the students' attention on several important aspects of the problem that she thought many students were ignoring. She attempted to help students move away from trial-and-error toward an analytic approach; students needed to think carefully about the different effects of reflecting and rotating a triangle. The class, *as a whole,* solved the problem quite analytically. But the teacher recognized that many individual students may not have completely understood what the class accomplished. So she had the students review these ideas individually by directing them to return to Shape Maker manipulations. This encouraged students to personally make sense of the previously discussed ideas. In the last part of the discussion, the teacher used Alex's question to reemphasize the importance of identifying a shape by examining its measurements and properties rather than its visual characteristics. Because of her knowledge of the van Hiele levels, she immediately understood the significance of Alex's comment and capitalized on this golden opportunity.

Proving It's a Square; Proving It's a Parallelogram

Proving It's a Square. The teacher begins by handing out a copy of a screen from the sketch Pairs--Right triangles (see Figure TE 4.8). The printout shows a right triangle, its reflection about its hypotenuse, and all the relevant measurements. She also projects a copy of the screen on an overhead projector.

Length(AC) = 112 pixels   Angle(BCA) = 45°
Length(AB) = 112 pixels   Angle(BAC) = 90°
Length(BC) = 158 pixels   Angle(ABC) = 45°

Length(A'C) = 112 pixels
Length(A'B) = 112 pixels
Angle(BCA') = 45°
Angle(BA'C) = 90°
Angle(A'BC) = 45°

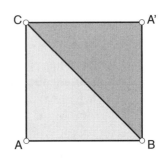

▲ Show reflection in AC
△ Hide reflection in AC
▲ Show reflection in AB
△ Hide reflection in AB
▲ Show reflection in BC
△ Hide reflection in BC

▲ Show rotation in AC
△ Hide rotation in AC
▲ Show rotation in AB
△ Hide rotation in AB
▲ Show rotation in BC
△ Hide rotation in BC

Figure TE 4.8

Teacher: Yesterday, we had this on the screen and somebody said that the figure was a square. How can we prove that this is a square?

Arnie: Angle BAC is 90°.

Teacher: Let's look at the diagram to see what angle Arnie is talking about. He said angle BAC. Do you see what angle he is talking about? Do you believe that's 90°? How can you prove that it is 90°?

Arnie: At the top, in the box, it says angle BAC equals 90°.

Teacher: Do you agree? Notice the position of the A. The middle letter A tells the vertex of the angle, where the two sides come together.

That's part of the proof, Arnie, but it's not the whole thing, is it?
Rick, would you like to add something?

Rick: Angle BA'C equals 90°.

Teacher: Do you have any proof that it is true?

Rick:	It says it up there, angle BA'C.
Steve:	All of the side lengths are equal.
Teacher:	Do you have any proof of that?
Steve:	AC is 112. C to A' is 112. The other two are 112.
Juan:	C to B to A' is 45°.
Teacher:	Is that a characteristic of a square?
Juan:	The angle of the triangle is 45°, and if you mirrored it, it has to be 90°. It's half of 90°. It is half of that angle. And that angle is 45°.
Teacher:	And how do you know that this has to be 45°?
Juan:	Because I looked in the blue box.
Tina:	Kind of adding on to Juan's, if you took out the line from C to B, the middle line, all the angles would be 90°. That line is there because there are two triangles.
Teacher:	Toshi, is there anything else that proves that it is a square?
Toshi:	Two sets of parallel lines. [These students had not studied the angle relationships necessary for parallelism.]
Teacher:	Dave, do you have anything you want to add?
Dave:	It is a square because there are four 90° angles. And from the last unit that we did in the Shape Makers, I knew that a square was 360°. And the way you can prove it is that a triangle is 180°. And a mirrored image of it would be another 180°. That equals 360°. And another way is there are four 90° angles and 90 times 4 equals 360.
Laura:	Well, I agree with what you said with parallel lines. But when you have four 90° angles, you might not have to say that. Because when you have four 90° angles you know that you have two sets of parallel lines.
Teacher:	So you are saying that we might not have had to list that opposite sides are parallel, but you would agree that it is a characteristic?
Laura:	Yeah, I agree!
Alex:	There should be four lines of symmetry.
Teacher:	Could you tell me where those lines of symmetry are?
Alex:	We can already see the first line of symmetry [pointing to B and C].
Teacher:	So if I took the square and folded it from here to here [B to C], one side would fold exactly onto the other side?
Alex:	Right! And you could fold it from A' to A. And from the middle of A' and C to the middle of B and A.
Teacher:	You want me to find the midpoint of A'C [unobtrusively adding a standard mathematical term to Alex's idea]?
Alex:	Yeah!
Teacher:	And where do I go with that?
Alex:	Just go straight down to the middle of A and B. And then from the middle of C to A to the middle of A' to B.

Analysis. In this episode, students are attempting to prove that a figure is a square. They are quite familiar with the properties of a square; they mention that it has 90° angles, two sets of parallel sides, all equal side lengths, and four lines of symmetry. They are able to *prove* that the sides are equal and that the angles are 90° by appropriately referring to measurements. But these fifth graders have not gone into enough detail to analytically justify their claims about parallelism and symmetry. The next example shows how students in another class were able to deal analytically with parallelism.

Proving It's a Parallelogram. In another class, students are discussing how they know that the shape formed by rotating a right triangle about a side is a parallelogram.

Teacher: For those who said yes, how do we know that you got a parallelogram?

Jason: Because it has two sets of parallel lines.

Teacher: How do you know that they are parallel?

Jason: Well, we wrote down our information because we thought somebody might question us about this. We rotated about side AB. We said that AC' and CB were parallel because the alternate interior angles—C'AB and ABC—were both 37°. We remembered when we did parallel stuff with quadrilaterals that we said that if these two angles were equal, then the lines were parallel. We showed that the other two lines were parallel the same way.

Teacher: Okay. Does everybody accept their argument?

Students: Yeah.

Teacher: Did anybody say anything else to prove that this is a parallelogram?

Sarah: Opposite sides are equal. We checked the pixel measurements on the screen.

Teacher: Anything else?

Brenda: Opposite angles are the same. We checked the angle measurements on the screen. Two of them were already up there. But the other pair, you had to add two angles to get the angles for the parallelogram. But they added up to be the same.

Teacher: Okay, you've given three properties of parallelograms—we talked about these a while ago when we studied quadrilaterals. I'm curious. Do you think we have to show that all three of these properties are true?

Mia: Yes. Because if it's a parallelogram, all of these things must be true.

Brandon: I'm not so sure. I think that if the opposite sides are parallel, then they must also be equal. You just can't make a quadrilateral with opposite sides parallel but not equal, because if they are not equal, then they can't stay parallel.

Jon: I think Brandon's right. But I also think that if you have opposite sides equal, then they will be parallel. Because if you make them equal, they stay even, so they are parallel.

Mia: Well, I see what they are saying, but I still think you have to show all three things, because that's what we said a parallelogram was. It's the only way to be sure.

Teacher: Okay, great discussion. All of you should keep thinking about these ideas. After we've thought about them some more, we'll talk about them again.

Analysis. Some of these students are starting to complete their transition to van Hiele level 3. They are starting to believe that one property implies another property, so to classify a shape you don't have to prove that it has all the properties for the class of shapes into which you would like to place it. It would be profitable to let these students discuss definitions for classes of shapes such as parallelograms. What should we take as the definition of a parallelogram if we want to make the definition as brief as possible—that is, to make it contain as few properties as possible? Of course, not all students may be ready for this jump in sophistication. Mia, for example, still believes that you need to prove all the properties; she might be quite unwilling to accept minimal definitions.

Another important point that must be mentioned here is the informal nature of students' reasoning. Even though Brandon and Jon believe that one property implies another, they are using informal reasoning to make their conclusions; they are thinking about what is possible based on mental models developed through their experiences with the Shape Makers. We could help them sharpen these mental models by asking them to test some of their ideas explicitly. For instance, when they use the Quadrilateral Maker to make a quadrilateral with opposite sides equal, are the opposite sides parallel? When they use the Quadrilateral Maker to make a quadrilateral with opposite sides parallel, are the opposite sides equal? However, a truly deductive proof of these implications is not yet available to these students, because they are not explicitly working in an axiomatic system. That will come in a later geometry course.

Tessellations

Summary

Students reflect or rotate a triangle about one of its sides to form a quadrilateral or a triangle. They then tessellate the plane with this shape and describe how they made their tessellation.

Mathematical Objectives

This exploration concludes students' investigation of quadrilateral and triangle Shape Makers on a fun note. In it, students investigate the art-related topic of tessellations from an analytic-mathematical perspective that enables them to describe the structure of the tessellations they create.

Mac:

📁 Triangle Makers

    📁 Triangle Exploration 5

        ◈ Triangle Tessellations

📁 Demonstration sketches

    📁 for Triangles

        ◈ Quad Tessellation Demo

Windows:

📁 TriMakrs

    📁 Tri_Exp5

        ◈ Tri_Tess.gsp

📁 Demos

    📁 Triangle

        ◈ QuadTess.gsp

Required Materials

Session	Student Sheet	SS#	Geometer's Sketchpad sketch
1 and 2	Triangle Tessellations	65, 66	Triangle Tessellations (Mac) Tri_Tess.gsp (Windows)

Make an overhead transparency of the Tessellation Transparency on page 224. You can use it to describe what a tessellation is.

Activity: Triangle Tessellations

Use Student Sheets 65 and 66, and sketch:

Mac: Windows:

      Triangle Exploration 5   Tri_Exp5

            Triangle Tessellations   Tri_Tess.gsp

Class Discussion

➡ Distribute the Triangle Tessellations student sheet to the class. Explain that the students' task is to make and describe a tessellation.

➡ Describe what a tessellation is:

A tessellation is a covering of the plane by copies of a shape, with no gaps or overlaps.

Use the Tessellation Transparency to show students several sample tessellations. Note that in some tessellations there may be several shapes that can be thought of as the tessellating shape. For instance, for the tessellation at the bottom right of the sheet, either the hexagon or the curved shape within it can be thought of as the tessellating shape:

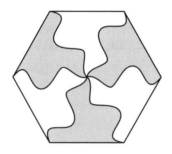

➡ Illustrate to students how to use the sketch Triangle Tessellations (as shown in the following drawing):

(a) Use the control points on the measured Triangle Maker to construct a triangle. (b) Construct a quadrilateral by double clicking on either the Show quadrilateral by reflection button or the Show quadrilateral by rotation button. (c) Double click on the appropriate buttons to display stage 1, the first circular layer of the tessellation, then stage 2, the second circular layer of the tessellation. You can still drag the control points on the triangle to change the tessellation.

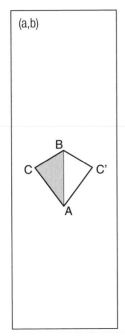

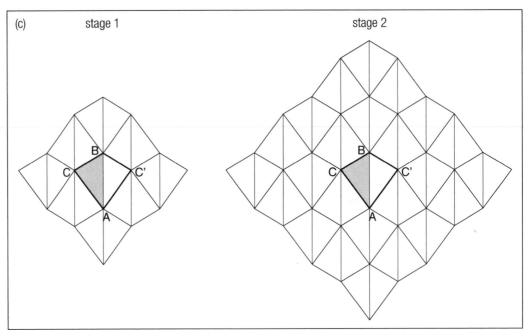

(a,b)

(c)　　stage 1　　　　　　　　　　　　stage 2

➡ As an optional demonstration, you can use the sketch Quad Tessellation Demo to illustrate how stage 1 of the quadrilateral tessellation is formed by successively rotating the quadrilateral about one of its sides. Drag the control point at the top of R1 clockwise around the blue semicircle to see the first rotation. Use a similar motion for rotations 2–8 in order. To return the rotated quadrilaterals to their original hidden position, double click on the Show button. Drag the magenta strip that connects R1–R8 counterclockwise until the control points return to their original positions. Double click on the Hide button to hide this strip.

Students Work in Pairs

➡ Have students experiment with making tessellations by manipulating the control points on the original (yellow) Triangle Maker. After creating a tessellation design they like, students record its vital statistics, print it, then describe its mathematical and artistic characteristics.

Students record on their student sheets the measurements of the triangle and quadrilateral formed, along with whether they reflected or rotated the triangle. (Note that the measurements for the quadrilateral are not given but must be derived from those of the triangle.)

Students print and color their tessellation. Before printing, students should double click on the Hide triangle labels button to remove the A, B, and C labels from the original triangle, then on the white triangle button, Δ, to hide all the other buttons. Double clicking on the solid triangle button, ▲, makes the other buttons reappear.

To conserve paper, you can restrict student tessellation prints to one page. To do this, before students print, have them choose **Print Preview** from the **File** menu, click on **Scale To Fit Page**, then click on **OK**.

Finally, students write a verbal description of their tessellation. The first sentence in this description should tell about the mathematical structure of their tessellation. It should be constructed in the form shown below. Other sentences should describe why students designed and colored the tessellation the way they did and what they see when they look at their tessellation.

First sentence in description:

My tessellation consists of a triangle that is $\left\{\begin{array}{c} scalene \\ isosceles \\ equilateral \end{array}\right\}$ and $\left\{\begin{array}{c} acute \\ obtuse \\ right \end{array}\right\}$

and has been $\left\{\begin{array}{c} reflected \\ rotated \end{array}\right\}$ about one of its sides to form a $\left\{\begin{array}{c} triangle \\ quadrilateral \\ square \\ rectangle \\ parallelogram \\ kite \\ rhombus \end{array}\right\}$.

If time permits, students can make several tessellations. Be sure they choose **Print Preview** before printing their tessellations so they don't waste paper printing designs that they don't like or that don't fit on a single sheet of paper.

Class Discussion

➡ After students have completed their tessellations, have them show and describe their creations to the class.

Students should describe the essential mathematical elements of their designs, including (a) type of triangle, (b) type of motion (reflection or rotation), and (c) type of figure formed (what kind of triangle or quadrilateral). They should justify the claims they made in the description by referring to appropriate properties and recorded measurements. Students should also describe what they see in their creations: "Mine looks like a bird." "Mine looks like steps." See "Teaching Note: Describing Our Tessellations" on page 136 for an illustration of how the discussion might proceed.

➡ Display students' tessellations and descriptions on a bulletin board. Students might decide how to group the tessellations by analyzing a combination of their geometric components and their design features.

Extensions

➡ Have students investigate differences between tessellations formed by reflecting a triangle about one of its sides and those formed by rotating it about one of its sides.

Students might notice, for instance, that rotating a triangle about one of its sides gives a parallelogram, so tessellations obtained by rotating the triangle all have the basic parallelogram structure shown here:

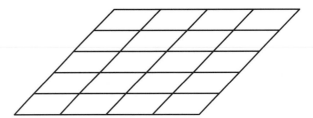

On the other hand, when you reflect a triangle about one of its sides, you get a kite or triangle, so tessellations obtained by reflecting the triangle have either a triangle structure or a kite structure:

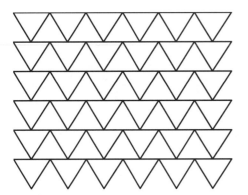

triangle structure

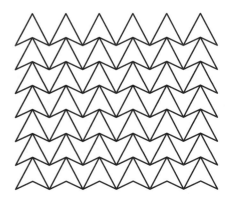

kite structure

➡ Have students investigate the following question: For what kinds of triangles is the tessellation obtained by reflecting the triangle about one of its sides the same as the tessellation obtained by rotating the triangle about one of its sides? Why? Students should make predictions, then check their predictions using the sketch Triangle Tessellations.

TEACHING NOTE

Describing Our Tessellations

Students are discussing their tessellations as a class. One by one, groups of students go to the front of the classroom to show and describe their tessellations.

[Alex and Steve come to the front of the room.]

Teacher: Show us your tessellation, point out and describe your original triangle, and describe what happened when you rotated or reflected it.

Alex: This is our original triangle [pointing to it on their printout].

Teacher: Can you describe it to us?

Steve: [Reading from his paper] It's scalene obtuse.

Teacher: Does anybody doubt that it's scalene obtuse [Alex points to the design]? What's one question you would like to ask to prove it's obtuse?

Jackie: What were the angles?

Teacher: Does she need all the angles?

[Some students shake their head no.]

Laura: They should give the obtuse angle.

Steve: One hundred and thirty-one.

Teacher: What else did you say about your triangle?

Alex: It's scalene.

Teacher: What do we need for that, one, two, or three pieces of information?

Steve: The side lengths are 82 pixels, 13 pixels, and 42 pixels.

Teacher: Does anybody disagree that it's scalene? Okay. So we have a scalene obtuse triangle. And what did you do next, a rotation or a reflection?

Alex: Reflection.

Teacher: What do you think they got for a new shape?

Brandy: Kite.

Teacher: Why?

Brandy: Because almost all the time when you reflect you get a kite, unless you reflect a certain way and then you get a triangle.

Teacher: Did you get a kite?

Alex: Yeah!

Teacher: And how did you know it was a kite?

Alex: Because the adjacent sides were equal [pointing appropriately to their quadrilateral].

Steve: One line of symmetry.

Alex: Four lines and four angles.

Teacher: Okay! Is there anybody who thinks this is not a kite?

Alex: Oh yeah, it's also like one set of opposite angles are equal. Like here and here [pointing to their quadrilateral] the angles are equal.

Toshi: But doesn't a kite have to have a line of symmetry from one vertex to the other vertex? If you didn't say that, we could've thought it was a square.

Teacher: Could what you have described be a square? What part of your description would've eliminated a square?

Alex: Well, I said one set of opposite angles are equal.

Steve: Not all of them.

Teacher: Would that help you?

Toshi: Yeah!

Alex: I also said that it doesn't have to have 90° angles.

[Sara, Ada, and Alexis come to the front of the room.]

Alexis: Our tessellation consists of a triangle that is scalene and obtuse and has been reflected about one of its sides to form a kite.

Teacher: Any questions for them?

Laura: What is the obtuse angle?

Alexis: It was 158°.

Teacher: Other questions for them?

Brandy: What were the measurements of the sides?

Teacher: Why are you asking?

Brandy: Because if it's a scalene, they all have to be different.

Teacher: Okay, what were the measurements?

Ada: We had 64, 222, and 161.

Teacher: So does that prove that it's scalene?

Students: Yeah.

[Mark and Arnie come to the front of the room.]

Arnie: [Reading from his sheet] Our tessellation consists of a triangle that is scalene and obtuse and it has been reflected about one of its sides to form a kite. Length AC is 94 pixels, length CB is 179 pixels, length BA is 109 pixels. And then angle C is 30°, angle B is 26°, and angle A is 124°. It is a kite because it has a mirrored image and the two adjacent sides are equal in length.

Teacher: What was your proof that it's a kite?

Arnie: Because it had the two adjacent sides equal: AC and AC′ equal 94; BC and BC′ equal 179. And there was a mirrored image and there was one line of symmetry.

Teacher: Does anybody disagree with him? [Nobody does.]

[Rick and Jamie come to the front of the room.]

Rick: This is our shape [shows design]. It's a triangle that's scalene obtuse. And we reflected on one of its sides to form a kite.

Teacher: Okay! Do you want to be drilled, or do you want to give us more information?

Rick: [After conferring with Jamie] We know it's a kite because it has one line of symmetry. It has a mirrored image, and the adjacent sides are equal.

Teacher: What else do you know? What did you call the triangle?

Rick: An obtuse triangle.

Teacher: Brandy.

Brandy: You said it was an obtuse triangle, but what about the side lengths?

Jamie: Well, all three side lengths are different.

Teacher: So what would we call that?

Rick: Scalene.

Teacher: Do we need to hear the actual side lengths since he told us they were all three different?

Students: No!

Teacher: Any other questions for them?

George: What's your obtuse angle?

Rick: Angle C is 153°.

Teacher: Any other questions? [No.]

[Naomi and Li Chen come to the front of the room.]

Naomi: We formed a scalene obtuse triangle. We reflected it to make a kite [shows the tessellation].

Teacher: Can you give us proofs?

Naomi: The line lengths for the scalene triangle are all different, 245 pixels, 166 pixels, and 84 pixels.

Li Chen: The angles are 8°, 17°, and 155°.

Teacher: What did you get?

Girls: A kite.

Teacher: How did you know that?

Li Chen: Because it had a mirrored image. And we knew that the Quadrilateral Maker and the Kite Maker were the only two Quadrilateral Makers that could make a V shape.

Teacher: Could you give me a more mathematical reason than just saying it could make a V?

Naomi: Okay. There are two adjacent sides that are equal, a mirrored image.

Li Chen: And there's a line of symmetry from vertex to vertex.

Teacher: Okay. Those are great.

[Dave and Tom come to the front of the room.]

Dave: Well, our tessellation consists of a triangle that is scalene and right. And it has been reflected about one of its sides to form an isosceles triangle. And we know that it's scalene because the angles [pointing at information on the design], angle C is 56°, angle B is 90°, and angle A is 34°. We know that it's right because angle B is 90°. It has a 90° angle, so it's a right triangle.

Teacher: Do you have a question, Alexis?

Alexis: When he said that it was scalene, he was pointing to the angles. He didn't give the side lengths.

Teacher: Okay. So why is it scalene, Dave?

Dave: Well, because the line length of AC is 36 pixels, the line length of CB is 20 pixels, and the line length of BA is 30 pixels. So none of the pixels are the same, so it has three different line lengths.

Teacher: And you reflected it and got a what?

Dave: An isosceles triangle.

Teacher: How do you know it's an isosceles?

Dave: We knew that it's isosceles because it has a line of symmetry [points to the design]. It has a mirrored image. It's a three-sided polygon.

[George and Harrison go to the front of the class.]

George: Our first shape was a scalene obtuse triangle. We know it's scalene because the lengths are 184, 123, and 76. And we know it's obtuse because one of the angles is 134°. We reflected it on one of its sides and it became a kite. We know that because it has the adjacent side lengths equal and it has one line of symmetry.

[Brandy and Laura come to the front of the room.]

Laura: [Brandy holds the design.] It's a triangle that is scalene obtuse because none of the side lengths are equal. And, it has one obtuse angle. And it's a kite [pointing to the design].

Brandy: When we reflected it, it became a kite because it has a mirror image. And the two sets of adjacent lines are the same. When we colored it [Brandy shows the colored design], we colored it so that we were looking at it as if it was steps. And we colored the part that you would step on blue. And the part that was in front we colored purple.

Teacher: Almost an optical illusion, isn't it?

Brandy: Yeah!

[Toshi goes to the front of the room.]

Toshi: My tessellation consists of a triangle that is scalene and obtuse and has been reflected about one of its sides to form a kite. I know it's a scalene because the side lengths are 192, 55, and 144. I know it's an obtuse because the obtuse angle is 147. I know it's a kite because it can't be a square, because it doesn't have four 90° angles. It can't be a rectangle because it doesn't have four 90° angles. It can't be a rhombus because all the sides aren't the same. It could be a quadrilateral, but it would be more descriptive to say it's a kite.

Teacher: Can you tell us more about why it is a kite?

Toshi: Because it has the two adjacent sides equal. It has a line of symmetry from one vertex to another vertex. And I think that's all it really needs to have.

Teacher: Any questions?

Brandy: How did you color it?

Toshi: I didn't do it any specific way [shows colored design]. I just colored it. Actually, I sort of thought it looked like a bird.

Analysis. As you can see from the examples, the class discussion of students' tessellations is not merely a "show-and-tell" session. Instead, the teacher keeps the discussion focused on a theme that runs through the whole *Shape Makers* unit—more precise analysis requires examination of shapes in terms of their properties rather than their holistic visual characteristics. Thus, this fun activity helps students return again to this fundamental geometric concept.

Extensions

You can take many possible directions in extending the Shape Maker activities already presented. Several of these directions are described here. These extensions can be implemented after the Quadrilateral and Triangle Explorations have been completed. (All sketches and folders referred to in this section are contained in the folder Extensions.)

Polygon Flats with Verbal Descriptions

Make up Polygon Flats mysteries in which only verbal clues are given and suspects include both quadrilateral and triangle Shape Makers. See Student Sheet 67, Polygon Flats Revisited, for an example.

Mystery Shape Makers Again

Students attempt to identify/describe the Mystery Shape Makers in the folder Mystery Shape Maker. There are ten quadrilateral Shape Makers in this folder, including the original seven quadrilateral Shape Makers and three new ones. Angle and length measurements are shown for each.

If students claim that a Mystery Shape Maker is one of the original seven, encourage them to prove their claim by referring to measurements and properties. If they claim that a Mystery Shape Maker is new, they should try to describe its identifying properties.

Note that this activity could be used for an assessment of students' understanding of the quadrilateral Shape Makers.

Minimal Definitions for Shapes

When students are in level 2 in the van Hiele hierarchy, they tend to list all the properties they can think of for a shape. As they move into level 3, they begin to see that one property can imply another; they begin to be capable of understanding and devising definitions that are not logically redundant. Once students get to the point of thinking that one property of a shape implies another—as illustrated in the section "Proving It's a Parallelogram" in "Teaching Note: Proving It's a Square; Proving It's a Parallelogram" on page 128—they can be asked to compose minimal definitions for shapes. For instance, in that teaching note, the teacher started students thinking about this issue when she asked, "Do you think we have to show that all three of these properties are true?" After exploring this question as suggested in the "Analysis" section of that teaching note, students could be asked: "What do you think we could use as a definition for *parallelogram*? What is the most concise definition we could use? That is, what definition lists the fewest properties?"

Logically, students might conclude, for instance, that "a parallelogram is a quadrilateral with opposite sides parallel." Or they might conclude that "a parallelogram is a quadrilateral with opposite sides equal." But their sense of clarity will probably cause them to be more attracted to the first of these two definitions. That is, although both definitions are minimal, the first is more pleasing aesthetically because it uses the concept of parallelism in defining a parallelogram.

Exploring Diagonals of Quadrilaterals

Students can profitably explore the diagonals of quadrilaterals. There are various ways to do this.

1. Sometimes students naturally start thinking about diagonals. For instance, after exploring the quadrilateral Shape Makers, fifth-grader Elizabeth made this conjecture and argument:

<u>Elizabeth's conjecture:</u>

> *The only time a Kite Maker and Rectangle Maker make the same shape is when they make a square.*

<u>Elizabeth's argument:</u>

> *Both the Kite Maker and the Rectangle Maker will make squares.*
>
> *The Kite Maker will not make rectangles because the diagonals in the Kite Maker always make right angles.*
>
> *But the only time the Rectangle Maker's diagonals make right angles is when you make a square.*

You can have students use the sketches in the folder Quads with Diagonals to investigate Elizabeth's conjecture. Double clicking on the Show diagonals button in these Shape Makers shows the diagonals and the measurements needed to determine how the diagonals are related.

Length(AB) = 72 pixels   Angle(A) = 110°
Length(BC) = 147 pixels   Angle(B) = 101°
Length(CD) = 147 pixels   Angle(C) = 47°
Length(AD) = 72 pixels   Angle(D) = 101°
Angle(AXD) = 90°
Distance(X to A) = 41 pixels
Distance(X to D) = 59 pixels
Distance(X to C) = 135 pixels
Distance(X to B) = 59 pixels

▲ Show possible parallels | ▲ Show possible symmetry line
△ Hide possible parallels | △ Hide possible symmetry line

▲ Show possible diagonals
△ Hide possible diagonals

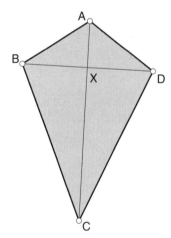

2. After students have explored a conjecture such as Elizabeth's, you can have them use the sketches in the folder Quads with Diagonals to investigate the diagonals of all six special quadrilateral Shape Makers. What can they conclude? They should seek to make statements such as Elizabeth did—"The diagonals in the Kite Maker always make right angles"—about each special quadrilateral Shape Maker.

3. You can give students riddles involving diagonals:

"My diagonals always bisect each other."

"My diagonals are always perpendicular."

Students solve the riddles using the sketches in the folder Quads with Diagonals. Inform students whether these riddles have more than one or only one solution.

Exploring Similarity

On each of the special Shape Makers in the folder Similarity are two red control points that change the size of the figures but not their shape. (The shape can be changed using the other vertex control points or the special green control points.) Have students manipulate these control points and investigate how the resulting figures are related. They should discover that these control points, although they change the size and perhaps the orientation of a figure, maintain the "shape" of the figure. They also keep the angles the same and corresponding sides proportional.

To provide students with an appropriate mental model for similarity, you can use the sketch Dilation Demo in the folder Similarity. Drag the red control point on the thick top segment in the upper-left corner to change the ratio of dilation. Show students that triangle A'B'C' is the same size as triangle ABC when the dilation ratio is 1. Students can find the ratios of corresponding side lengths (e.g., A'B'/AB) to see that they are all the same as the dilation ratio.

Note that X is the center of dilation. Double click on the Show rays button to see how the rays XAA', XBB', and XCC' "project out" from X. Showing these rays gives a nice visual illustration of how triangle ABC is projected outward to form triangle A'B'C'. Students can even think of these rays as "rays of light" emanating from the point source of light, X.

Inscribing Quadrilaterals and Triangles in Circles

Investigating which triangles and quadrilaterals can be inscribed in a circle is a nice exploration. For example, the triangle and quadrilateral shown here are inscribed in circles—all their vertices lie on the circle. (Another way of describing the relationship between these polygons and the circles is to say that the polygons are *circumscribed* by the circles.)

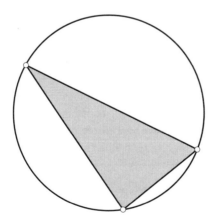
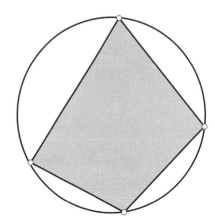

To start students in this investigation, first show them what it means to inscribe a triangle and a quadrilateral in a circle. Then ask them to use the sketches in the folder Polygons in Circles to investigate which kinds of triangles and quadrilaterals can be inscribed in a circle. Each sketch contains a measured triangle or quadrilateral Shape Maker, along with a circle in which you can attempt to inscribe it.

Once students have concluded that some quadrilaterals can be inscribed in a circle and some cannot (all triangles can), have students try to determine a criterion for inscribability. Students can pursue this question using the sketch Quadrilateral and Circle. They should find that when a quadrilateral can be inscribed in a circle, its opposite angles are supplementary.

One way to see why this is true is to explore the sketch Angles in Circle. As shown here, in this sketch the central angle has its vertex at the center of the circle; the inscribed angle has its vertex on the circle. Students should find that the measure of the inscribed angle is half the measure of the intercepted arc, DE, which has the same measure as the central angle.

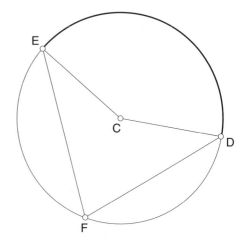

Central angle(DCE) = 148°
Arc(DE) = 148°
Inscribed angle(DFE) = 74°

Thus, if a quadrilateral is inscribed in a circle, each pair of opposite angles intercepts a pair of arcs that make up the whole circle. Because the two arcs in a pair have a total measure of 360° and an inscribed angle is half the measure of its intercepted arc, the two opposite angles have a total measure of 180°.

Additional insight into this problem can be gained by exploring the two sketches Quadrilateral Maker with PBs and Triangle Maker with PBs in the folder With Perpendicular Bisectors. In these two sketches, the polygons have perpendicular bisectors for each side, along with circles in which you can attempt to inscribe them. Students can note that the perpendicular bisectors of the three sides of a triangle always intersect at a single point (the center of the circumscribing circle), while the perpendicular bisectors of the four sides of a quadrilateral intersect at a single point only when the vertices of the quadrilateral all lie on a circle. The common point of intersection of the four perpendicular bisectors of a quadrilateral is the center of the circumscribing circle.

To understand the role of perpendicular bisectors in these deliberations, students can explore the sketch Perpendicular Bisector. In it, they can see that every point on the perpendicular bisector is equidistant from the two endpoints of the segment. This idea can be used to show why every triangle can be inscribed in a circle.

For instance, in the triangle below, because line DK is the perpendicular bisector of segment AC, every point on it is equidistant from points A and C. Because line FK is the perpendicular bisector of segment AB, every point on it is equidistant from points A and B. And because point K is on both DK and FK, it is equidistant from A, B, and C. Thus, these three points lie on a circle centered at K.

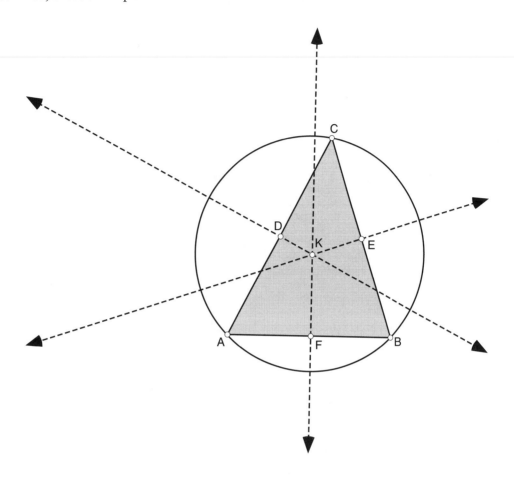

This is a difficult argument for most junior high students to follow. It is presented here only to suggest how you might attempt to guide students' investigations so that they can invent it on their own.

Students Construct Their Own Shape Makers

Finally, after students have been introduced to some of the construction capabilities of The Geometer's Sketchpad, you can challenge them to create their own Shape Makers. Students can attempt to make their own versions of Shape Makers that already exist, or they might try to make some not included with the *Shape Makers* software. For instance, some students have attempted to make an Acute Triangle Maker.

The construction of some Shape Makers is straightforward; others are more difficult. You might observe students making a certain type of special Shape Maker, say a Trapezoid Maker, and find that their creation makes only a proper subset of all trapezoids. This is still quite an accomplishment, and, if it happens, students should be encouraged to describe the class of shapes that their Shape Maker makes. If other students have made the same Shape Maker, students can compare both the operation and the construction of their products. You might also observe student creations with extra control points—ones not at the vertices—as in the sketches Similar Quadrilateral Maker and Similar Triangle Maker in the folder Similarity.

Other Shape Makers that students might attempt are polygons that have more than four sides. Students could also attempt to construct special polygons with more than four sides. For instance, can they make an Equilateral or Equiangular Pentagon, Hexagon, or Octagon Maker? With these types of Shape Makers, they can investigate the question "For what *n* does an equilateral polygon have to be equiangular, or vice versa?"

Geometric Glossary

Acute angle—An acute angle is an angle that measures between 0° and 90°.

Acute triangle—An acute triangle has all of its interior angles acute.

Angle—An angle is the union of two rays with a common endpoint.

Circle—A circle is the set of all points in a plane that are a fixed distance away from a given point. The given point is called the *center*. Any line segment from its center to the circle itself is called a *radius*.

Concave polygon—A polygon is concave if at least one of its diagonals passes through the exterior of the polygon.

Congruent—Two *shapes* are congruent if they have the exact same shape and size, that is, if one can be placed exactly on top of the other. Two *line segments* are congruent if they have the same length. Two *angles* are congruent if they have the same measure. Two *polygons* are congruent if their corresponding sides and angles are congruent.

Convex polygon—A polygon is convex if all of its diagonals lie in its interior.

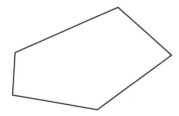

Diagonal of a quadrilateral—A diagonal of a polygon is a line segment between two nonadjacent vertices.

Equal—Two geometric objects are equal if they consist of exactly the same points.

Equilateral triangle—An equilateral triangle has all of its sides congruent.

Exterior angle—An exterior angle of a polygon is formed by one side of the polygon and the line that is the extension of an adjacent side, like angle b below.

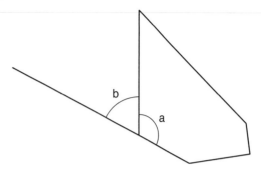

Hypotenuse—The hypotenuse of a right triangle is the side opposite the right angle.

Interior angle—An interior angle of a polygon is formed by two adjacent sides of the polygon and is in the inside or interior of the polygon, like angle a in the polygon above.

Intersect—Two lines intersect if they have at least one point in common.

Isosceles triangle—An isosceles triangle has at least two sides congruent.

Kite—A kite is a quadrilateral in which at least one diagonal defines a line of symmetry.

Line—A line is a continuous set of points that extends straight and indefinitely in two opposite directions.

Line segment—A line segment is the set of all points between two points on a line.

Midpoint—The midpoint of a line segment is the point on the segment that is equidistant from the endpoints of the segment.

Obtuse angle—An obtuse angle is an angle that measures between 90° and 180°.

Obtuse triangle—An obtuse triangle has one interior angle that is obtuse.

Parallel—Two lines in a plane are parallel if they do not intersect. Two line segments in a plane are parallel if the lines containing them are parallel.

Parallelogram—A parallelogram is a quadrilateral that has opposite sides parallel.

Perpendicular—Two lines are perpendicular if the angle between them is a right angle. Two line segments are perpendicular if the lines through them are perpendicular.

Perpendicular bisector—The perpendicular bisector of a line segment is the line that is perpendicular to the segment and passes through the segment's midpoint.

Pixel—A pixel is a small square dot on the computer screen.

Plane—A plane is a flat surface that extends indefinitely in all directions on the surface.

Point—A point is a location in space.

Polygon—A polygon is a simple, closed curve consisting of line segments only.

Quadrilateral—A quadrilateral is a four-sided polygon.

Rectangle—A rectangle is a quadrilateral that has opposite sides congruent and four right angles.

Reflection—In a plane, a reflection is a transformation in which a shape is transformed into its mirror image. The position of the line about which the shape is transformed (called the *line of reflection* or *flip line*) determines the position of the image.

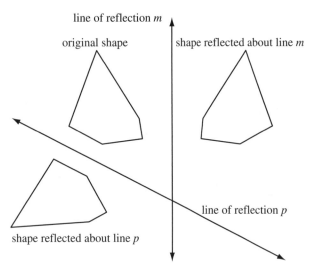

Rhombus—A rhombus is a quadrilateral that has all sides congruent.

Right angle—A right angle measures 90°.

Right triangle—A right triangle has one interior angle that is a right angle.

Rotation—In a plane, a rotation is a transformation in which a shape is turned about a point. The point is called the *center of rotation*. This center and the amount of turn determine the position of the image. In the figure below, the original shape has been rotated 90° about point X.

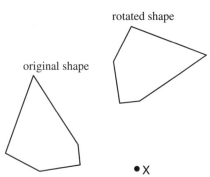

Scalene triangle—A scalene triangle has no sides congruent.

Side—A side of a polygon is one of its component line segments.

Similar—Two *shapes* are similar if they have the exact same shape but not necessarily the same size. Two *polygons* are similar if their corresponding angles are congruent and their corresponding sides are proportional.

Simple closed curve—A simple closed curve is a curve that starts and ends at the same point and that does not intersect itself anywhere except at this start/end point.

Square—A square is a quadrilateral that has all sides congruent and four right angles.

Symmetry—A figure has line symmetry or is symmetric about a line if its reflection, or flip, about that line is equal to the figure (that is, lands exactly on top of the original figure).

Tessellation— A tessellation is a covering of the plane by copies of a shape, with no gaps or overlaps.

Trapezoid—In this book, a trapezoid is a quadrilateral that has at least one set of opposite sides parallel. (In some books, trapezoids are defined as having one and only one set of opposite sides parallel.)

Triangle—A triangle is a three-sided polygon.

Vertex—A vertex of a polygon is the point of intersection of two of its sides. *Vertices* (the plural of *vertex*) occur at the endpoints of the sides of the polygon.

STUDENT SHEETS

MAKE YOUR OWN PICTURE

Mac:

📁 Quadrilateral Exploration 1

◈ Quadrilateral Makers

Windows:

📁 QuadExp1

◈ Quads.gsp

Use all seven of the quadrilateral Shape Makers to make a picture. Draw your picture below. Label the part of the picture made by each Shape Maker.

Which Shape Maker can make the *least* number of different types of shapes?

_____

Why? _____

_____

_____

Which Shape Maker can make the *greatest* number of different types of shapes?

_____

Why? _____

_____

_____

CAN YOU MAKE IT? #1

Mac: Windows:

📁 Quadrilateral Exploration 1   📁 QuadExp1

◆ Can You Make It? #1   ◆ CYMI_1.gsp

Use all seven quadrilateral Shape Makers to make the picture below.

Tell which Shape Maker you used to make each shape and why you used it.

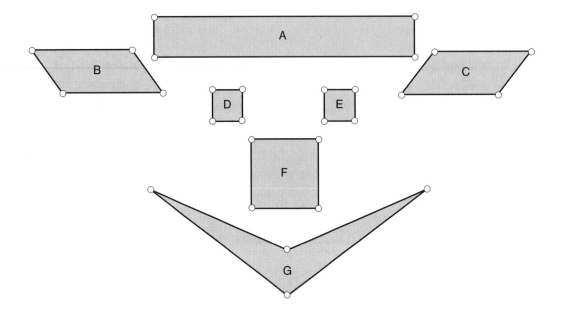

Shape	Shape Maker	Why?
A		
B		
C		
D		
E		
F		
G		

CAN YOU MAKE IT? #2

Mac:

Quadrilateral Exploration 1

Can You Make It? #2

Windows:

QuadExp1

CYMI_2.gsp

Use all seven quadrilateral Shape Makers to make the picture at right.

Tell which Shape Maker you used to make each shape and why you used it.

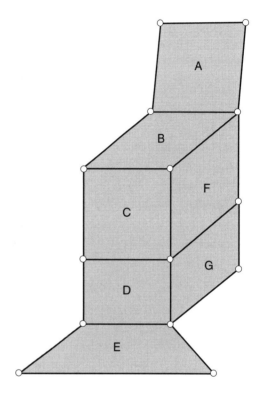

Shape	Shape Maker	Why?
A		
B		
C		
D		
E		
F		
G		

SHAPE MAKER CHALLENGE GAME (page 1 of 3)

Mac:

Quadrilateral Exploration 2

Quadrilateral Makers

Windows:

QuadExp2

Quads.gsp

Rules and Directions

The Shape Maker Challenge Game is for two players. The player who goes first in a game is called Player 1. (Partners should take turns going first in games.) To start a game, Player 1 uses the Shape Maker Selector to randomly choose the Shape Maker he or she will use for this game, then writes his or her initials next to that Shape Maker in the Player 1 column for Game 1 on the student sheet. Player 2 then uses the Shape Maker Selector to randomly choose a Shape Maker different from Player 1. Player 2 puts his or her initials next to the selected Shape Maker in the Player 2 column for Game 1.

To play, Player 1 makes a shape with his or her Shape Maker that he or she thinks Player 2 can't make. Player 2 then tries to make this shape with his or her Shape Maker. If Player 2 successfully makes Player 1's shape, then Player 2 tries to make a shape that Player 1 can't make. The players keep taking turns until one player cannot make the shape made by the other player. The player who makes a shape that his or her opponent cannot make is the winner.

After a game is won, players circle the initials of the winner, then draw the shape that could not be made and explain why it couldn't be made.

To ensure that the shapes made can be clearly seen, players must follow the size rule: (Part 1) The shapes that players make with their Shape Makers must be small enough to fit completely on the screen. (Part 2) The shapes must be large enough that the circle formed by the flat end of a standard pencil eraser will fit inside.

SHAPE MAKER CHALLENGE GAME (page 2 of 3)

Game	Player 1	Player 2	Draw the shape that couldn't be made.	Why couldn't this shape be made?
1	Square Maker Rectangle Maker Parallelogram Maker Rhombus Maker Trapezoid Maker Kite Maker	Square Maker Rectangle Maker Parallelogram Maker Rhombus Maker Trapezoid Maker Kite Maker		
2	Square Maker Rectangle Maker Parallelogram Maker Rhombus Maker Trapezoid Maker Kite Maker	Square Maker Rectangle Maker Parallelogram Maker Rhombus Maker Trapezoid Maker Kite Maker		
3	Square Maker Rectangle Maker Parallelogram Maker Rhombus Maker Trapezoid Maker Kite Maker	Square Maker Rectangle Maker Parallelogram Maker Rhombus Maker Trapezoid Maker Kite Maker		

SHAPE MAKER CHALLENGE GAME (page 3 of 3)

Game	Player 1	Player 2	Draw the shape that couldn't be made.	Why couldn't this shape be made?
4	Square Maker Rectangle Maker Parallelogram Maker Rhombus Maker Trapezoid Maker Kite Maker	Square Maker Rectangle Maker Parallelogram Maker Rhombus Maker Trapezoid Maker Kite Maker		
5	Square Maker Rectangle Maker Parallelogram Maker Rhombus Maker Trapezoid Maker Kite Maker	Square Maker Rectangle Maker Parallelogram Maker Rhombus Maker Trapezoid Maker Kite Maker		
6	Square Maker Rectangle Maker Parallelogram Maker Rhombus Maker Trapezoid Maker Kite Maker	Square Maker Rectangle Maker Parallelogram Maker Rhombus Maker Trapezoid Maker Kite Maker		

IDENTIFY THE HIDING SHAPE MAKERS 1

Mac:

☐ Quadrilateral Exploration 2

◈ Hiding Shape Makers 1

Windows:

☐ QuadExp2

◈ Hide_SM1.gsp

Square Maker	**Trapezoid Maker**
Rectangle Maker	**Kite Maker**
Parallelogram Maker	**Quadrilateral Maker**
Rhombus Maker	

The Shape Makers listed above have had their name tags replaced with letters so you can't tell which is which.

Figure out the real name for each Shape Maker.

Describe how you figured out the true identity of each Shape Maker.

Letter	Real name	How did you figure it out?
A		
B		
C		
D		
E		
F		
G		

PREDICT AND CHECK (page 1 of 3)

Mac:

Quadrilateral Exploration 3

P&C Square Maker, etc.

Windows:

QuadExp3

PC_Squar.gsp, etc.

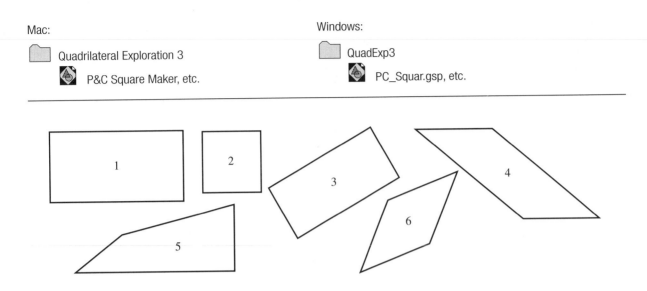

For each *measured* quadrilateral Shape Maker:

    a. Predict which of the above shapes the Shape Maker can make.

    b. Check your predictions with the P&C Shape Maker sketches.

Circle your answers. If a Shape Maker cannot make a shape, explain why not.

Can this Shape Maker	Make this shape?	I predict:	I found:	I *found* that this Shape Maker will *not* make this shape because:
Measured Square Maker	1	Y N	Y N	
	2	Y N	Y N	
	3	Y N	Y N	
	4	Y N	Y N	
	5	Y N	Y N	
	6	Y N	Y N	

PREDICT AND CHECK (page 2 of 3)

Can this Shape Maker	Make this shape?	I predict:		I found:		I *found* that this Shape Maker will *not* make this shape because:
Measured Rectangle Maker	1	Y	N	Y	N	
	2	Y	N	Y	N	
	3	Y	N	Y	N	
	4	Y	N	Y	N	
	5	Y	N	Y	N	
	6	Y	N	Y	N	
Measured Parallelogram Maker	1	Y	N	Y	N	
	2	Y	N	Y	N	
	3	Y	N	Y	N	
	4	Y	N	Y	N	
	5	Y	N	Y	N	
	6	Y	N	Y	N	
Measured Rhombus Maker	1	Y	N	Y	N	
	2	Y	N	Y	N	
	3	Y	N	Y	N	
	4	Y	N	Y	N	
	5	Y	N	Y	N	
	6	Y	N	Y	N	

Shape Makers ©1998 by Key Curriculum Press |

PREDICT AND CHECK (page 3 of 3)

Can this Shape Maker	Make this shape?	I predict:		I found:		I *found* that this Shape Maker will *not* make this shape because:
Measured Trapezoid Maker	1	Y	N	Y	N	
	2	Y	N	Y	N	
	3	Y	N	Y	N	
	4	Y	N	Y	N	
	5	Y	N	Y	N	
	6	Y	N	Y	N	
Measured Kite Maker	1	Y	N	Y	N	
	2	Y	N	Y	N	
	3	Y	N	Y	N	
	4	Y	N	Y	N	
	5	Y	N	Y	N	
	6	Y	N	Y	N	
Measured Quadrilateral Maker	1	Y	N	Y	N	
	2	Y	N	Y	N	
	3	Y	N	Y	N	
	4	Y	N	Y	N	
	5	Y	N	Y	N	
	6	Y	N	Y	N	

IDENTIFY THE HIDING SHAPE MAKERS 2

Mac:

📁 Quadrilateral Exploration 3

🔷 Hiding Shape Makers 2

Windows:

📁 QuadExp3

🔷 Hide_SM2.gsp

Square Maker	**Trapezoid Maker**
Rectangle Maker	**Kite Maker**
Parallelogram Maker	**Quadrilateral Maker**
Rhombus Maker	

The Shape Makers listed above have had their name tags replaced with letters so you can't tell which is which.

Figure out the real name for each Shape Maker.

Give as many reasons as you can to justify that each of your answers is correct.

Letter	Real name	Justifications
A		
B		
C		
D		
E		
F		
G		

WHICH SHAPE MAKERS CAN YOU USE? (page 1 of 3)

Mac:

📁 Quadrilateral Exploration 4

    ◈ M Square Maker, etc.

Windows:

📁 QuadExp4

    ◈ M_Square.gsp, etc.

For each *measured* quadrilateral Shape Maker:

    a. Predict whether or not the Shape Maker can make each quadrilateral described in 1–6.

    b. Check your predictions with the M Shape Maker sketches.

Circle your answers. If a Shape Maker cannot make one of the described quadrilaterals, explain why not.

    1. A quadrilateral with side lengths 20, 30, 40, 50.

    2. A quadrilateral with side lengths 30, 30, 50, 50.

    3. A quadrilateral with all angles different.

    4. A quadrilateral with angles 45°, 135°, 45°, 135°.

    5. A quadrilateral with two pairs of sides parallel.

    6. A quadrilateral with *at least* one line of symmetry.

Can this Shape Maker	Make this shape?	I predict:		I found:		I *found* that this Shape Maker will *not* make this shape because:
Measured Square Maker	1	Y	N	Y	N	
	2	Y	N	Y	N	
	3	Y	N	Y	N	
	4	Y	N	Y	N	
	5	Y	N	Y	N	
	6	Y	N	Y	N	

WHICH SHAPE MAKERS CAN YOU USE? (page 2 of 3)

Can this Shape Maker	Make this shape?	I predict:		I found:		I *found* that this Shape Maker will *not* make this shape because:
Measured Rectangle Maker	1	Y	N	Y	N	
	2	Y	N	Y	N	
	3	Y	N	Y	N	
	4	Y	N	Y	N	
	5	Y	N	Y	N	
	6	Y	N	Y	N	
Measured Parallelogram Maker	1	Y	N	Y	N	
	2	Y	N	Y	N	
	3	Y	N	Y	N	
	4	Y	N	Y	N	
	5	Y	N	Y	N	
	6	Y	N	Y	N	
Measured Rhombus Maker	1	Y	N	Y	N	
	2	Y	N	Y	N	
	3	Y	N	Y	N	
	4	Y	N	Y	N	
	5	Y	N	Y	N	
	6	Y	N	Y	N	

Shape Makers ©1998 by Key Curriculum Press |

WHICH SHAPE MAKERS CAN YOU USE? (page 3 of 3)

Can this Shape Maker	Make this shape?	I predict:		I found:		I *found* that this Shape Maker will *not* make this shape because:
Measured Trapezoid Maker	1	Y	N	Y	N	
	2	Y	N	Y	N	
	3	Y	N	Y	N	
	4	Y	N	Y	N	
	5	Y	N	Y	N	
	6	Y	N	Y	N	
Measured Kite Maker	1	Y	N	Y	N	
	2	Y	N	Y	N	
	3	Y	N	Y	N	
	4	Y	N	Y	N	
	5	Y	N	Y	N	
	6	Y	N	Y	N	
Measured Quadrilateral Maker	1	Y	N	Y	N	
	2	Y	N	Y	N	
	3	Y	N	Y	N	
	4	Y	N	Y	N	
	5	Y	N	Y	N	
	6	Y	N	Y	N	

HOW ARE THEY THE SAME? (page 1 of 2)

Mac:

📁 Quadrilateral Exploration 4

    ◈ M Square Maker, etc.

Windows:

📁 QuadExp4

    ◈ M_Square.gsp, etc.

To check your answers, use the measured quadrilateral Shape Makers.

1. Describe everything that is the same about all shapes made with the Square Maker.

_____

_____

_____

_____

_____

2. Describe everything that is the same about all shapes made with the Rectangle Maker.

_____

_____

_____

_____

3. Describe everything that is the same about all shapes made with the Parallelogram Maker.

_____

_____

_____

_____

HOW ARE THEY THE SAME? (page 2 of 2)

4. Describe everything that is the same about all shapes made with the Rhombus Maker.

_____

_____

_____

_____

_____

5. Describe everything that is the same about all shapes made with the Trapezoid Maker.

_____

_____

_____

_____

_____

6. Describe everything that is the same about all shapes made with the Kite Maker.

_____

_____

_____

_____

_____

SHAPE MAKER RIDDLES: WHO AM I? (page 1 of 5)

Mac:

📁 Quadrilateral Exploration 4

◆ M Quadrilateral Maker, etc.

Windows:

📁 QuadExp4

◆ M_Quad.gsp, etc.

Pretend that the Shape Makers can talk and that they like to pose riddles about themselves.

First, PREDICT which *one* Shape Maker is described by a riddle, and explain the reasoning you used to convince yourself that your prediction is correct.

Second, CHECK your answer with the measured Shape Makers, and explain the reasoning you used to convince yourself that your check is correct.

---

1. No matter how I change, I always have two pairs of sides with the same length and at least one line of symmetry. Sometimes I have only one line of symmetry. Which Shape Maker am I?

Prediction _____

Explain the reasoning you used in making your prediction.

_____

_____

_____

Check _____

Explain the reasoning you used for your check.

_____

_____

_____

STUDENT
SHEET **18**

SHAPE MAKER RIDDLES: WHO AM I? (page 2 of 5)

> 2. **No matter how I change, I always have at least one pair of parallel sides. Which Shape Maker am I?**

Prediction _____

Explain the reasoning you used in making your prediction.

_____

_____

_____

Check _____

Explain the reasoning you used for your check.

_____

_____

_____

> 3. **Whenever I have a right angle, I'm a rectangle. Sometimes not all of my sides have the same length and I have no right angles. Which Shape Maker am I?**

Prediction _____

Explain the reasoning you used in making your prediction.

_____

_____

_____

Check _____

Explain the reasoning you used for your check.

_____

_____

_____

170 | **Shape Makers** ©1998 by Key Curriculum Press

SHAPE MAKER RIDDLES: WHO AM I? (page 3 of 5)

> 4. I'm always a parallelogram. I'm always a kite. Sometimes I'm not all right. Which Shape Maker am I?

Prediction _____

Explain the reasoning you used in making your prediction.

_____

_____

_____

Check _____

Explain the reasoning you used for your check.

_____

_____

_____

> 5. I always have at least two lines of symmetry. Sometimes not all of my sides have the same length. Which Shape Maker am I?

Prediction _____

Explain the reasoning you used in making your prediction.

_____

_____

_____

Check _____

Explain the reasoning you used for your check.

_____

_____

_____

SHAPE MAKER RIDDLES: WHO AM I? (page 4 of 5)

---

6. **I can have two right angles or four, but never just one. Which Shape Maker am I?**

Prediction _____

Explain the reasoning you used in making your prediction.

_____

_____

_____

Check _____

Explain the reasoning you used for your check.

_____

_____

_____

---

7. **I always have the measures of any two adjacent angles sum to 180°. Sometimes not all of my sides have the same length. I don't always have right angles. Which Shape Maker am I?**

Prediction _____

Explain the reasoning you used in making your prediction.

_____

_____

_____

Check _____

Explain the reasoning you used for your check.

_____

_____

_____

SHAPE MAKER RIDDLES: WHO AM I? (page 5 of 5)

8. Sometimes I have four different side lengths. Sometimes no pair of my angles has measures that sum to 180°. Which Shape Maker am I?

Prediction _____

Explain the reasoning you used in making your prediction.

_____

_____

_____

Check _____

Explain the reasoning you used for your check.

_____

_____

_____

ANGLES IN INTERSECTING AND PARALLEL LINES (page 1 of 3)

Mac:

📁 Quadrilateral Exploration 4

◈ Lines and Angles

Windows:

📁 QuadExp4

◈ Line_Ang.gsp

1. Drag the control point on line *k* to make several different values for angles A and B. Record the values in the table below.

   a. How are the measures of these angles related?

   _____

   _____

   _____

   _____

   _____

Angle A	Angle B

   b. Why do you think this is true?

   _____

   _____

   _____

2. Double click on the Show Supplements of A & B button. Drag the control point on line *k* to make several different values for angles A, B, C, and D. Record the values in the table below.

Angle A	Angle B	Angle C	Angle D

ANGLES IN INTERSECTING AND PARALLEL LINES (page 2 of 3)

a. How are the measures of these angles related?

_____

_____

_____

b. Why do you think this is true?

_____

_____

_____

3. Double click on the Show Line *n* button. Drag the control point on line *n* to make lines *m* and *n* parallel. To decide if the lines are parallel, drag line *n* so that the arcs on it are on top of the arcs on line *m*—if the lines are parallel, line *n* will fit on top of line *m*.

When lines *m* and *n* are parallel, describe the relationships that exist between angles A, B, C, D and angles E, F, G, H.

_____

_____

_____

_____

_____

Describe why you think these relationships occur.

_____

_____

_____

_____

ANGLES IN INTERSECTING AND PARALLEL LINES (page 3 of 3)

Are these relationships still true if lines *m* and *n* are not parallel?

_____

Explain why or why not.

_____

_____

_____

_____

4. In the diagram below, line *a* is parallel to line *b*. Angle 1 has measure 100°. Without using measuring instruments, find the exact measures of angles 2–8. Explain how you found your answers.

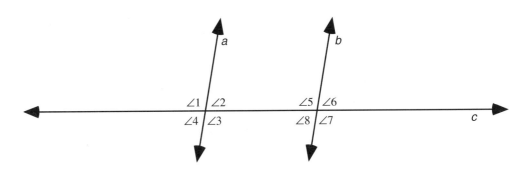

Angle	Measure	How did you find the angle measure?
2		
3		
4		
5		
6		
7		
8		

THE MYSTERY OF POLYGON FLATS #1 (page 1 of 2)

Mac: Windows:

📁 Quadrilateral Exploration 5/6   📁 QuadEx56

  ◈ M Square Maker, etc.   ◈ M_Square.gsp, etc.

Famous female detective Shirley Lock-Holmes is in the two-dimensional town of Polygon Flats investigating a theft in Quadrilateral Mansion. The seven people who live in the mansion are listed below. These people can change their shape and size, but only to a shape that can be made by their Shape Maker.

Sudha Square (played by the Square Maker)	**Trapezoid Tracy** (played by the Trapezoid Maker)
Rectangle Rick (played by the Rectangle Maker)	**Ricardo Rhombus** (played by the Rhombus Maker)
Kaneisha Kite (played by the Kite Maker)	**Quentin Quadrilateral** (played by the Quadrilateral Maker)
Parallelogram Pete (played by the Parallelogram Maker)	

When the theft occurred, the people in the mansion looked as follows in the mansion's security cameras. Examine the evidence, including the camera pictures and clue below, and figure out who committed the theft.

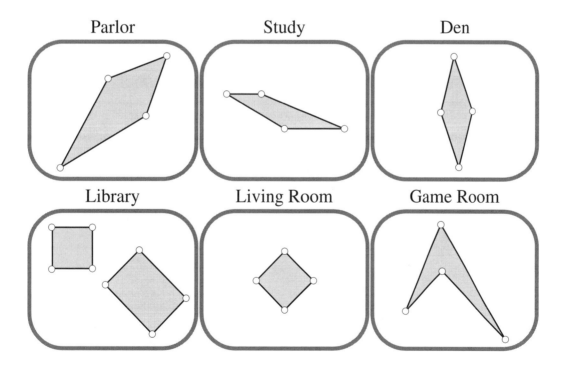

Parlor	Study	Den

Library	Living Room	Game Room

THE MYSTERY OF POLYGON FLATS #1 (page 2 of 2)

Clue
1. The theft occurred in the study.

Who is the thief? _____

Write an argument that proves "beyond the shadow of doubt" that the person you have accused of being the thief is guilty.

_____

_____

_____

_____

_____

_____

THE MYSTERY OF POLYGON FLATS #2 (page 1 of 2)

Mac:

📁 Quadrilateral Exploration 5/6

◇ M Square Maker, etc.

Windows:

📁 QuadEx56

◇ M_Square.gsp, etc.

Famous detective Shirley Lock-Holmes is in the two-dimensional town of Polygon Flats investigating a theft in Quadrilateral Mansion. The seven people who live in the mansion are listed below. These people can change their shape and size, but only to a shape that can be made by their Shape Maker.

Sudha Square (played by the Square Maker)
Rectangle Rick (played by the Rectangle Maker)
Kaneisha Kite (played by the Kite Maker)
Parallelogram Pete (played by the Parallelogram Maker)

Trapezoid Tracy (played by the Trapezoid Maker)
Ricardo Rhombus (played by the Rhombus Maker)
Quentin Quadrilateral (played by the Quadrilateral Maker)

When the theft occurred, the people in the mansion looked as follows in the mansion's security cameras. Examine the evidence, including the camera pictures and clues below, and figure out who committed the theft.

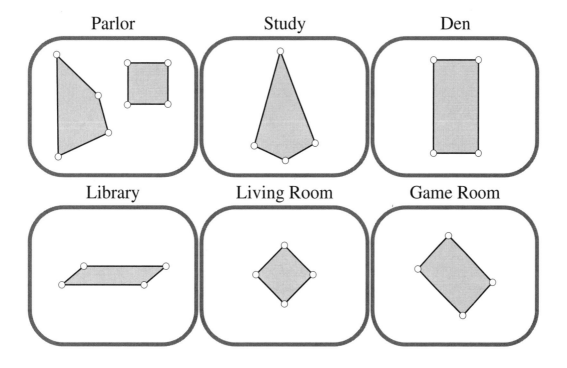

Parlor   Study   Den

Library   Living Room   Game Room

THE MYSTERY OF POLYGON FLATS #2 (page 2 of 2)

Clues
1. **The theft occurred in the library.**
2. **The thief sometimes has four unequal sides.**

Who is the thief? _____

Write an argument that proves "beyond the shadow of doubt" that the person you have accused of being the thief is guilty.

_____

_____

_____

_____

_____

_____

THE MYSTERY OF POLYGON FLATS #3 (page 1 of 2)

Mac:

📁 Quadrilateral Exploration 5/6

🔷 M Square Maker, etc.

Windows:

📁 QuadEx56

🔷 M_Square.gsp, etc.

Famous detective Shirley Lock-Holmes is in the two-dimensional town of Polygon Flats investigating a theft in Quadrilateral Mansion. The seven people who live in the mansion are listed below. These people can change their shape and size, but only to a shape that can be made by their Shape Maker.

Sudha Square (played by the Square Maker)	**Trapezoid Tracy** (played by the Trapezoid Maker)
Rectangle Rick (played by the Rectangle Maker)	**Ricardo Rhombus** (played by the Rhombus Maker)
Kaneisha Kite (played by the Kite Maker)	**Quentin Quadrilateral** (played by the Quadrilateral Maker)
Parallelogram Pete (played by the Parallelogram Maker)	

When the theft occurred, the people in the mansion looked as follows in the mansion's security cameras. Examine the evidence, including the camera pictures and clues below, and figure out who committed the theft.

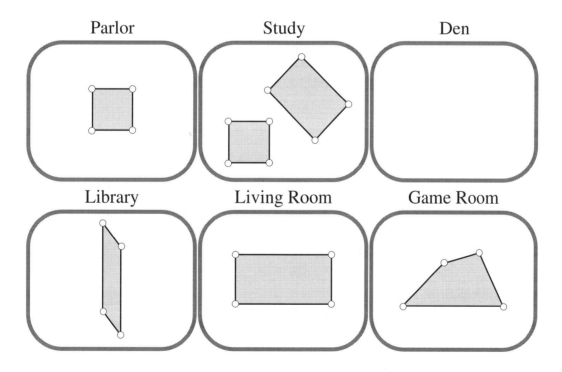

Parlor   Study   Den

Library   Living Room   Game Room

THE MYSTERY OF POLYGON FLATS #3 (page 2 of 2)

Clues
1. The theft occurred in the parlor. 2. People who are always right are always in the same room. 3. A well-balanced person (who always has at least two lines of symmetry) was out of the mansion at the time of the theft.

Who is the thief? _____

Write an argument that proves "beyond the shadow of doubt" that the person you have accused of being the thief is guilty.

_____

_____

_____

_____

_____

_____

THE MYSTERY OF POLYGON FLATS #4 (page 1 of 2)

Mac:

📁 Quadrilateral Exploration 5/6

🔷 M Square Maker, etc.

Windows:

📁 QuadEx56

🔷 M_Square.gsp, etc.

Famous detective Shirley Lock-Holmes is in the two-dimensional town of Polygon Flats investigating a theft in Quadrilateral Mansion. The seven people who live in the mansion are listed below. These people can change their shape and size, but only to a shape that can be made by their Shape Maker.

Sudha Square (played by the Square Maker)	**Trapezoid Tracy** (played by the Trapezoid Maker)
Rectangle Rick (played by the Rectangle Maker)	**Ricardo Rhombus** (played by the Rhombus Maker)
Kaneisha Kite (played by the Kite Maker)	**Quentin Quadrilateral** (played by the Quadrilateral Maker)
Parallelogram Pete (played by the Parallelogram Maker)	

When the theft occurred, the people in the mansion looked as follows in the mansion's security cameras. Examine the evidence, including the camera pictures and clues below, and figure out who committed the theft.

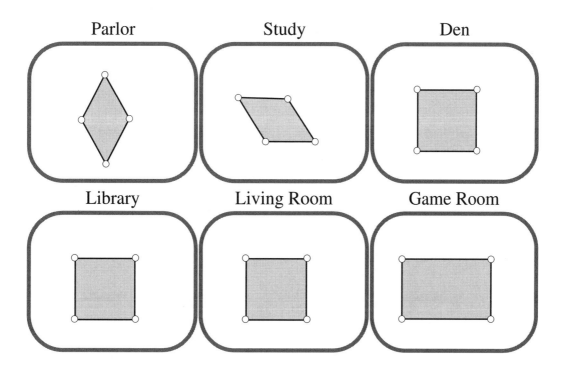

Parlor   Study   Den

Library   Living Room   Game Room

THE MYSTERY OF POLYGON FLATS #4 (page 2 of 2)

Clues
1. The theft occurred in the game room.
2. People in the parlor and game room always have at least two equal sides.
3. At the time of the theft, one person was at the airport, unstrung because there was not enough wind to fly.
4. The person in the study always has at least two lines of symmetry but sometimes has exactly two.

Who is the thief? _____

Write an argument that proves "beyond the shadow of doubt" that the person you have accused of being the thief is guilty.

_____

_____

_____

_____

_____

_____

MAKE YOUR OWN MYSTERY OF POLYGON FLATS (page 1 of 4)

Mac:

Quadrilateral Exploration 5/6

Your Own Polygon Flats

Windows:

QuadEx56

Your_PF.gsp

You have been invited by the *Polygonal Press* to write a mystery puzzle for next week's newspaper. Because of your extensive experience with the Shape Makers and your logical reasoning abilities, the editors feel that you can write a Polygon Flats puzzle that will throw a curve to the weekly readers.

Your task is to write your own Polygon Flats Mystery.

Directions

1. Open the sketch Your Own Polygon Flats on the computer. On the screen, you will see the seven Shape Makers labeled with their names.

2. You must first decide which Shape Maker has committed the crime and in what room the crime was committed. You must then decide which room each of the other Shape Makers was in when the crime was committed. It is possible for two Shape Makers to be in the same room. On a piece of scrap paper, record the room location for each Shape Maker and who committed the theft.

3. Now you are ready to use the computer to move each Shape Maker into the room where you have decided it should go. To do this, move the Shape Maker into the room where it belongs. Now change the Shape Maker into a shape of your choosing. For example, if the Parallelogram Maker is in the study, you may want it to make a parallelogram that other Shape Makers could also make. If the Parallelogram Maker is in the shape of a rectangle, then the reader won't know whether the Rectangle Maker is in the study or the Parallelogram Maker is in the study. This will make the puzzle more difficult to solve.

4. You should now write clues that will help readers find out which Shape Maker committed the crime. Record the clues on page 2 of the Make Your Own Mystery of Polygon Flats student sheet. Also on this sheet, draw what the Shape Makers in each room look like (without their names, of course).

5. After you have completed all the details needed to solve the puzzle, you are ready to hide the names of each Shape Maker. To do this, double click on the Hide names button.

MAKE YOUR OWN MYSTERY OF POLYGON FLATS (page 2 of 4)

6. (Optional) Your teacher may want you to save your work. To do this, go to the **File** menu and choose **Save As**. A dialog box will appear asking you to save your puzzle. Name your sketch according to your teacher's instructions.

7. Now you are ready for a test run. Invite another pair of students to solve your puzzle. Have them come over to your computer. Show them the clues that you have written and your mystery computer screen. Have them try to solve your mystery. Keep your answer hidden. At this time they are not allowed to manipulate the Shape Makers on the screen. Silently observe how they solve your mystery.

8. When the students have solved your puzzle, they can check it on the screen. To do this, they double click on the Show names button.

Revising Your Mystery

1. The students who solve your mystery should explain to you their solution and the strategy they used to solve your mystery. If they cannot solve your mystery, they should explain why they were unable to do so.

2. When these students are finished solving your mystery, you will go to their computer and solve their mystery.

3. There may be a class discussion when all students have solved one mystery. This is an opportunity for any groups who did not agree on a solution to come to a resolution with the help of the rest of the class.

4. You may now go back to your computer and make revisions to your mystery. You may wish to make changes

   ■ if it was unsolvable by the other pair of students;

   ■ if your mystery did not make sense to them and you need to change some clues or shapes.

5. After making your revisions, find *another* pair of students to solve your mystery. You attempt to solve their mystery. Again, talk about the strategies you used and about any other revisions that might be needed. Was your first revision successful?

6. Make any further revisions that are needed.

7. Your class may wish to put together a book with all of the students' mysteries.

MAKE YOUR OWN MYSTERY OF POLYGON FLATS (page 3 of 4)

Mac:

☐ Quadrilateral Exploration 5/6

◇ Your Own Polygon Flats

Windows:

☐ QuadEx56

◇ Your_PF.gsp

Written by _____

Famous detective Shirley Lock-Holmes is in the two-dimensional town of Polygon Flats investigating a theft in Quadrilateral Mansion. The seven people who live in the mansion are listed below. These people can change their shape and size, but only to a shape that can be made by their Shape Maker.

Sudha Square (played by the Square Maker)	**Trapezoid Tracy** (played by the Trapezoid Maker)
Rectangle Rick (played by the Rectangle Maker)	**Ricardo Rhombus** (played by the Rhombus Maker)
Kaneisha Kite (played by the Kite Maker)	**Quentin Quadrilateral** (played by the Quadrilateral Maker)
Parallelogram Pete (played by the Parallelogram Maker)	

When the theft occurred, here's how the people in the mansion looked in the mansion's security cameras. Examine the evidence, including the drawings on this student sheet, the clues on the next student sheet, and the camera pictures on the computer. Figure out who committed the theft.

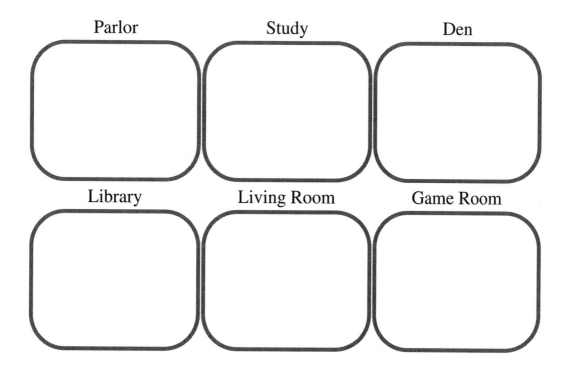

Parlor	Study	Den

Library	Living Room	Game Room

MAKE YOUR OWN MYSTERY OF POLYGON FLATS (page 4 of 4)

Clues

Who is the thief? _____

Write an argument that proves "beyond the shadow of doubt" that the person you have accused of being the thief is guilty.

_____

_____

_____

_____

_____

RELATING SHAPE MAKERS (page 1 of 4)

Mac:

📁 Quadrilateral Exploration 5/6

◈ M Square Maker, etc.

Windows:

📁 QuadEx56

◈ M_Square.gsp, etc.

> **1. Can every shape made by the Square Maker also be made by the Rectangle Maker? Explain why.**

Prediction _____

Explain the reasoning you used in making your prediction.

_____

_____

_____

Check _____

Explain the reasoning you used for your check.

_____

_____

_____

> **2. Can every shape made by the Rectangle Maker also be made by the Square Maker? Explain why.**

Prediction _____

Explain the reasoning you used in making your prediction.

_____

_____

_____

RELATING SHAPE MAKERS (page 2 of 4)

Check _____

Explain the reasoning you used for your check.

_____

_____

_____

3. Why can the Rectangle Maker be used to make squares?

_____

_____

_____

_____

_____

4. What property or rule could be added to the Rectangle Maker to make it a Square Maker? Explain why.

_____

_____

_____

_____

_____

RELATING SHAPE MAKERS (page 3 of 4)

> **5.** Can every shape made by the Parallelogram Maker also be made by the Rectangle Maker? Explain why.

Prediction _____

Explain the reasoning you used in making your prediction.

_____

_____

_____

Check _____

Explain the reasoning you used for your check.

_____

_____

_____

> **6.** Can every shape made by the Rectangle Maker also be made by the Parallelogram Maker? Explain why.

Prediction _____

Explain the reasoning you used in making your prediction.

_____

_____

_____

Check _____

Explain the reasoning you used for your check.

_____

_____

_____

RELATING SHAPE MAKERS (page 4 of 4)

7. Why can the Parallelogram Maker be used to make rectangles?

_____

_____

_____

_____

_____

8. What property or rule could be added to the Parallelogram Maker to make it a Rectangle Maker? Explain why.

_____

_____

_____

_____

_____

MATHEMATICAL DEBATES (page 1 of 2)

Mac:

☐ Quadrilateral Exploration 5/6

◈ M Parallelogram Maker, etc.

Windows:

☐ QuadEx56

◈ M_Para.gsp, etc.

Circle "True" or "False" for each statement. Then write an argument that will convince the class that your answer is correct.

1. Every rectangle is also a parallelogram.	True	False

Argument _____

_____

_____

_____

2. Every parallelogram is also a rectangle.	True	False

Argument _____

_____

_____

_____

3. Every rhombus is also a quadrilateral.	True	False

Argument _____

_____

_____

_____

MATHEMATICAL DEBATES (page 2 of 2)

| 4. Every quadrilateral is also a rhombus. | True | False |

Argument _____

_____

_____

_____

| 5. Every square is also a rectangle. | True | False |

Argument _____

_____

_____

_____

| 6. Every rectangle is also a square. | True | False |

Argument _____

_____

_____

_____

CAN YOU MAKE THE PICTURE? #1

Mac:

📁 Triangle Exploration 1

🔷 Can You Make the Picture? #1

Windows:

📁 Tri_Exp1

🔷 CYMP_1.gsp

Use each of the five triangle Shape Makers to make the picture below.

Tell which triangle Shape Maker you used to make each triangle and why.

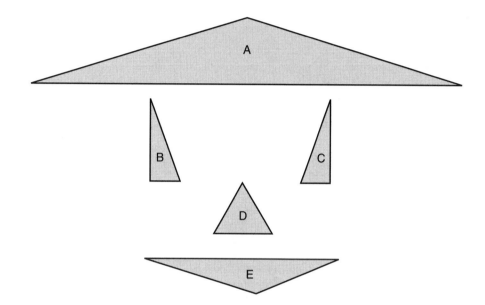

Shape	Shape Maker	Why?
A		
B		
C		
D		
E		

CAN YOU MAKE THE PICTURE? #2

Mac:

📁 Triangle Exploration 1

🔺 Can You Make the Picture? #2

Windows:

📁 Tri_Exp1

🔺 CYMP_2.gsp

Use each of the five triangle Shape Makers to make the picture below.

Tell which triangle Shape Maker you used to make each triangle and why.

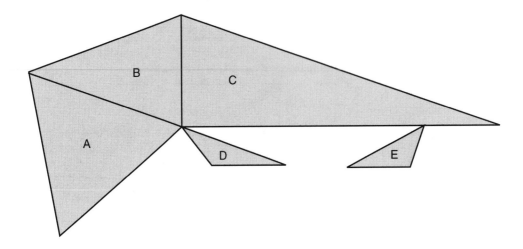

Shape	Shape Maker	Why?
A		
B		
C		
D		
E		

INVESTIGATING MEASURED TRIANGLE MAKERS (page 1 of 2)

Mac:

☐ Triangle Exploration 2

◈ M Equilateral Triangle Maker, etc.

Windows:

☐ Tri_Exp2

◈ M_Eqtri.gsp, etc.

---

1. Predict which measured Triangle Makers can make a triangle with *all of its sides equal*. Check your answers with the measured Triangle Makers.

Measured Triangle Maker	Predict:		Check:		Explain your prediction.
Isosceles	Y	N	Y	N	
Equilateral	Y	N	Y	N	
Right	Y	N	Y	N	
Obtuse	Y	N	Y	N	

What do you notice about triangles that have all sides equal?

_____

2. Predict which measured Triangle Makers can make a triangle with exactly *two 45° angles*. Check your answers with the measured Triangle Makers.

Measured Triangle Maker	Predict:		Check:		Explain your prediction.
Isosceles	Y	N	Y	N	
Equilateral	Y	N	Y	N	
Right	Y	N	Y	N	
Obtuse	Y	N	Y	N	

What do you notice about triangles that have two 45° angles?

_____

INVESTIGATING MEASURED TRIANGLE MAKERS (page 2 of 2)

3. Predict which measured Triangle Makers can make a triangle with *at least two equal angles*. Check your answers with the measured Triangle Makers.

Measured Triangle Maker	Predict:		Check:		Explain your prediction.
Isosceles	Y	N	Y	N	
Equilateral	Y	N	Y	N	
Right	Y	N	Y	N	
Obtuse	Y	N	Y	N	

What do you notice about triangles with at least two equal angles?

_____

4. Predict which measured Triangle Makers can make a triangle with *at least two sides equal*. Check your answers with the measured Triangle Makers.

Measured Triangle Maker	Predict:		Check:		Explain your prediction.
Isosceles	Y	N	Y	N	
Equilateral	Y	N	Y	N	
Right	Y	N	Y	N	
Obtuse	Y	N	Y	N	

What do you notice about triangles with at least two sides equal?

_____

THE SUM OF THE ANGLES IN A TRIANGLE

Mac:

    📁 Triangle Exploration 2

       🔷 M Triangle Maker

Windows:

    📁 Tri_Exp2

       🔷 M_Triang.gsp

Use the sketch M Triangle Maker to make three different triangles.

Record the measures of each triangle's three angles in the table below. Also record the *sum* of the three angles for each triangle.

Triangle	Angle measures	Sum of angle measures
1	Angle A =	
	Angle B =	
	Angle C =	
2	Angle A =	
	Angle B =	
	Angle C =	
3	Angle A =	
	Angle B =	
	Angle C =	

What did you find? Why do you think this is true?

_____

_____

_____

_____

_____

HOW ARE THE TRIANGLES THE SAME?

Mac:

☐ Triangle Exploration 2

◆ M Isosceles Triangle Maker, etc.

Windows:

☐ Tri_Exp2

◆ M_Isotri.gsp, etc.

Use the measured Triangle Makers to answer the following questions.

1. What's the same about all triangles made with the Isosceles Triangle Maker?

_____

_____

2. What's the same about all triangles made with the Equilateral Triangle Maker?

_____

_____

3. What's the same about all triangles made with the Right Triangle Maker?

_____

_____

4. What's the same about all triangles made with the Obtuse Triangle Maker?

_____

_____

OTHER KINDS OF TRIANGLES (page 1 of 6)

Mac:

📁 Triangle Exploration 3

◆ Other Kinds of Triangles 1

Windows:

📁 Tri_Exp3

◆ Ot_Tri1.gsp

Problem 1 What can we say about the side lengths of triangles that *cannot* be made by either the Equilateral Triangle Maker or the Isosceles Triangle Maker? Write your conjecture below.

Conjecture _____

_____

_____

_____

Experiment Use the blue Triangle Maker in the sketch Other Kinds of Triangles 1 to make three or four triangles that cannot be made with either the Equilateral Triangle Maker (green) or the Isosceles Triangle Maker (yellow).

For each triangle you make, use the table below to

1. draw the triangle

2. record the lengths of its sides

3. describe why it could not be made by either the Equilateral Triangle Maker or the Isosceles Triangle Maker

Drawing	Side lengths	Why can't it be made?

OTHER KINDS OF TRIANGLES (page 2 of 6)

Drawing	Side lengths	Why can't it be made?

Conclusion As a result of your experiment, what can you say about the side lengths of the triangles you drew?

_____

_____

_____

Explain why you think this must be true.

_____

_____

_____

_____

OTHER KINDS OF TRIANGLES (page 3 of 6)

Mac:

📁 Triangle Exploration 3

    ◈ Other Kinds of Triangles 2

Windows:

📁 Tri_Exp3

    ◈ Ot_Tri2.gsp

Problem 2 What can we say about the angles of triangles that cannot be made by either the Right Triangle Maker or the Obtuse Triangle Maker?

Conjecture _____

_____

_____

_____

Experiment Use the blue Triangle Maker in the sketch Other Kinds of Triangles 2 to make three or four triangles that cannot be made with either the Right Triangle Maker (yellow) or the Obtuse Triangle Maker (green).

For each triangle you make, use the table below to

1. draw the triangle
2. record the measures of its angles
3. describe why it could not be made by either the Right Triangle Maker or the Obtuse Triangle Maker

Drawing	Angle measures	Why can't it be made?

OTHER KINDS OF TRIANGLES (page 4 of 6)

Drawing	Angle measures	Why can't it be made?

Conclusion As a result of your experiment, what can you say about the angle measures of the triangles you drew?

_____

_____

_____

Explain why you think this must be true.

_____

_____

_____

_____

OTHER KINDS OF TRIANGLES (page 5 of 6)

Mac:

📁 Triangle Exploration 3

    ◆ Other Kinds of Triangles 1 & 2

Windows:

📁 Tri_Exp3

    ◆ Ot_Tri1.gsp, Ot_Tri2.gsp

Use the appropriate measured Triangle Makers to help you solve these problems.

Problem 3 What do you think an *isosceles right triangle* is?

_____

_____

_____

Can you make an isosceles right triangle with one of the Shape Makers?

_____

Explain how you made it.

_____

_____

_____

_____

Can you make an isosceles right triangle with a different Shape Maker?

_____

Explain.

_____

_____

_____

_____

_____

OTHER KINDS OF TRIANGLES (page 6 of 6)

How could you convince someone that the triangle you made should be called an isosceles right triangle?

_____

_____

_____

Problem 4 Can you make an *obtuse equilateral triangle?*

_____

Explain why or why not.

_____

_____

_____

_____

_____

MAKING SHAPES FROM TRIANGLE PAIRS (page 1 of 3)

Mac:

📁 Triangle Exploration 4

◈ Pairs--Triangles

Windows:

📁 Tri_Exp4

◈ Pr_Trian.gsp

Problem What kinds of shapes do you get when you reflect or rotate a triangle about one of its sides and combine the two triangles?

Experiment with the Triangle Maker in the sketch Pairs--Triangles. Move the control points to make different triangles. Double click on the Show and Hide buttons to see the shapes formed by reflecting or rotating the triangles.

1. Describe each different kind of shape you found (that is, say whether it is a quadrilateral, triangle, parallelogram, rhombus, rectangle, kite, square, and so on).

2. Circle whether you formed the shape by reflecting or rotating the triangle about one of its sides.

3. For each kind of shape you find, make the shape by cutting triangles out of paper. Tape the shape next to its description.

Describe the shape formed.	How was it formed? (Circle one.)	Tape two triangles together to make the shape.
	rotation reflection	

MAKING SHAPES FROM TRIANGLE PAIRS (page 2 of 3)

Describe the shape formed.	How was it formed? (Circle one.)	Tape two triangles together to make the shape.
	rotation reflection	
	rotation reflection	

MAKING SHAPES FROM TRIANGLE PAIRS (page 3 of 3)

Describe the shape formed.	How was it formed? (Circle one.)	Tape two triangles together to make the shape.
	rotation reflection	
	rotation reflection	

MAKING SPECIAL QUADRILATERALS FROM TRIANGLE PAIRS (page 1 of 5)

Mac:

📁 Triangle Exploration 4

    ◈ Pairs--Right triangles

    ◈ Pairs--Obtuse triangles

    ◈ Pairs--Isosceles triangles

    ◈ Pairs--Equilateral triangles

Windows:

📁 Tri_Exp4

    ◈ Pr_Rttri.gsp

    ◈ Pr_Obtri.gsp

    ◈ Pr_Istri.gsp

    ◈ Pr_Eqtri.gsp

For each type of Triangle Maker—Right, Obtuse, Isosceles, Equilateral—in the Pairs sketches, tell which special quadrilaterals can be made by rotating or reflecting the triangle about one of its sides.

PREDICT first (using drawing, paper cutting, or visualization).

Then CHECK with the Triangle Makers (Pairs--Right triangles and so on).

RECORD your findings in the tables on the following pages.

1. For predictions and checks, circle Y if the special quadrilateral can be made, N if it can't.

2. If your check shows that you can make a special quadrilateral, tell HOW you did it. Circle *rotate* or *reflect,* and circle which side you reflected or rotated about.

3. *After you have checked* whether a type of Triangle Maker can make a special quadrilateral:

    a. If you made the quadrilateral, EXPLAIN why you think it is what you say it is. For example, if you think you have formed a rectangle, how could you convince somebody that the shape is a rectangle?

    b. If you couldn't make the quadrilateral, EXPLAIN why it can't be made.

MAKING SPECIAL QUADRILATERALS FROM
TRIANGLE PAIRS (page 2 of 5)

Shape Maker	Predict:	Check:	How?	Explain why.
Right Triangle Maker	square Y N	square Y N	rotate or reflect about AB AC BC	
	rectangle that is not a square Y N	rectangle that is not a square Y N	rotate or reflect about AB AC BC	
	parallelogram that is not a rectangle Y N	parallelogram that is not a rectangle Y N	rotate or reflect about AB AC BC	
	kite that is not a square Y N	kite that is not a square Y N	rotate or reflect about AB AC BC	
	rhombus that is not a square Y N	rhombus that is not a square Y N	rotate or reflect about AB AC BC	

MAKING SPECIAL QUADRILATERALS FROM
TRIANGLE PAIRS (page 3 of 5)

Shape Maker	Predict:	Check:	How?	Explain why.
Obtuse Triangle Maker	square Y N	square Y N	rotate or reflect about AB AC BC	
	rectangle that is not a square Y N	rectangle that is not a square Y N	rotate or reflect about AB AC BC	
	parallelogram that is not a rectangle Y N	parallelogram that is not a rectangle Y N	rotate or reflect about AB AC BC	
	kite that is not a square Y N	kite that is not a square Y N	rotate or reflect about AB AC BC	
	rhombus that is not a square Y N	rhombus that is not a square Y N	rotate or reflect about AB AC BC	

MAKING SPECIAL QUADRILATERALS FROM
TRIANGLE PAIRS (page 4 of 5)

Shape Maker	Predict:	Check:	How?	Explain why.
Isosceles Triangle Maker	square Y N	square Y N	rotate or reflect about AB AC BC	
	rectangle that is not a square Y N	rectangle that is not a square Y N	rotate or reflect about AB AC BC	
	parallelogram that is not a rectangle Y N	parallelogram that is not a rectangle Y N	rotate or reflect about AB AC BC	
	kite that is not a square Y N	kite that is not a square Y N	rotate or reflect about AB AC BC	
	rhombus that is not a square Y N	rhombus that is not a square Y N	rotate or reflect about AB AC BC	

MAKING SPECIAL QUADRILATERALS FROM
TRIANGLE PAIRS (page 5 of 5)

Shape Maker	Predict:	Check:	How?	Explain why.
Equilateral Triangle Maker	square Y N	square Y N	rotate or reflect about AB AC BC	
	rectangle that is not a square Y N	rectangle that is not a square Y N	rotate or reflect about AB AC BC	
	parallelogram that is not a rectangle Y N	parallelogram that is not a rectangle Y N	rotate or reflect about AB AC BC	
	kite that is not a square Y N	kite that is not a square Y N	rotate or reflect about AB AC BC	
	rhombus that is not a square Y N	rhombus that is not a square Y N	rotate or reflect about AB AC BC	

WHAT KIND OF TRIANGLE? WHAT KIND OF QUADRILATERAL? (page 1 of 2)

Length(AB) = 118 pixels	Angle(BCA) = 71°
Length(AC) = 118 pixels	Angle(BAC) = 38°
Length(BC) = 76 pixels	Angle(ABC) = 71°

Length(AB') = 76 pixels
Length(B'C) = 118 pixels
Angle(B'CA) = 38°
Angle(B'AC) = 71°
Angle(AB'C) = 71°

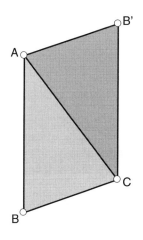

1. What kind of triangle is triangle ABC?

_____

2. How can you prove your answer correct?

_____

_____

_____

_____

WHAT KIND OF TRIANGLE? WHAT KIND OF QUADRILATERAL? (page 2 of 2)

3. Was triangle ABC rotated or reflected to make quadrilateral ABCB'?

_____

4. How can you prove your answer correct?

_____

_____

_____

_____

5. What kind of quadrilateral is shape ABCB'?

_____

6. How can you prove your answer correct?

_____

_____

_____

_____

TRIANGLE TESSELLATIONS (page 1 of 2)

Mac:

▢ Triangle Exploration 5

◈ Triangle Tessellations

Windows:

▢ Tri_Exp5

◈ Tri_Tess.gsp

1. Make any triangle you want by manipulating the control points of the yellow measured Triangle Maker. Form a quadrilateral by double clicking either the Show quadrilateral by reflection or the Show quadrilateral by rotation button.

2. Double click on the appropriate buttons for showing stage 1, then stage 2, of either the reflection or the rotation tessellation. Make different tessellations by manipulating the control points on the original yellow Triangle Maker.

3. When you have created a tessellation design that you like, record below the essential information about your triangle and quadrilateral.

My triangle was (check one):　　❏　reflected　　　❏　rotated

Measurements for my triangle	Measurements for my quadrilateral
Length AB = _____　　Angle A = _____	Length AC = _____　　Angle A = _____
Length AC = _____　　Angle B = _____	Length AC′ = _____　　Angle B = _____
Length BC = _____　　Angle C = _____	Length BC′ = _____　　Angle C = _____
	Length BC = _____　　Angle C′ = _____

4. Print and color your tessellation. (Be sure to hide the labels and buttons before you print.)

TRIANGLE TESSELLATIONS (page 2 of 2)

5. Describe the tessellation you printed and colored. The first sentence in your description should tell about the mathematical structure of your tessellation. It should be written in the form shown below. Other sentences should tell why you designed and colored the tessellation the way you did and what you see when you look at your tessellation.

First sentence in description:

My tessellation consists of a triangle that is $\left\langle \begin{array}{c} scalene \\ isosceles \\ equilateral \end{array} \right\rangle$ and $\left\langle \begin{array}{c} acute \\ obtuse \\ right \end{array} \right\rangle$

and has been $\left\langle \begin{array}{c} reflected \\ rotated \end{array} \right\rangle$ about one of its sides to form a $\left\langle \begin{array}{c} triangle \\ quadrilateral \\ square \\ rectangle \\ parallelogram \\ kite \\ rhombus \end{array} \right\rangle$.

Description:

POLYGON FLATS REVISITED

Mac:

📁 Quadrilateral Exploration 4

🔺 Triangle Exploration 2

Windows:

📁 QuadExp4

🔺 Tri_Exp2

Famous detective Shirley Lock-Holmes is in the two-dimensional town of Polygon Flats investigating a theft in Quadrilateral Mansion. There were 12 people staying at the mansion at the time of the theft. They are listed below. Figure out who committed the theft and prove your answer correct.

Sudha Square (played by the Square Maker)
Rectangle Rick (played by the Rectangle Maker)
Kaneisha Kite (played by the Kite Maker)
Parallelogram Pete (played by the Parallelogram Maker)
Trapezoid Tracy (played by the Trapezoid Maker)
Ricardo Rhombus (played by the Rhombus Maker)

Quentin Quadrilateral (played by the Quadrilateral Maker)
Isosceles Iris (played by the Isosceles Triangle Maker)
Obtuse Olaf (played by the Obtuse Triangle Maker)
Tina Triangle (played by the Triangle Maker)
Rheta Right (played by the Right Triangle Maker)
Equilateral Evan (played by the Equilateral Triangle Maker)

Clues

1. At the time of the theft, all people with at least one line of symmetry were outdoors practicing for their circus high-wire act.
2. The thief sometimes has all angles different.
3. The thief always has two or three angles whose measures sum to 180°.
4. The thief is not a female.

Who is the thief? _____

Write an argument that proves "beyond a shadow of a doubt" that the person you have accused of being the thief is guilty.

_____

_____

_____

_____

_____

Conjectures and Queries

Conjectures

Describe any ideas that you think might be true, even if you are not sure about them.

_____

_____

_____

_____

_____

_____

_____

Queries

Describe any questions you have about the Shape Makers or about what other students or the teacher has said about them.

_____

_____

_____

_____

_____

_____

_____

SHAPE MAKER SELECTORS

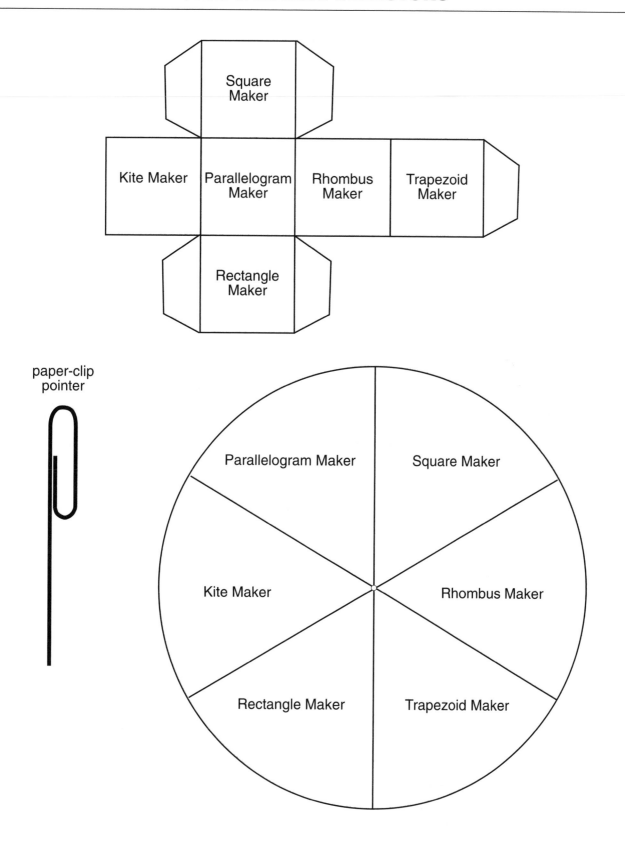

paper-clip
pointer

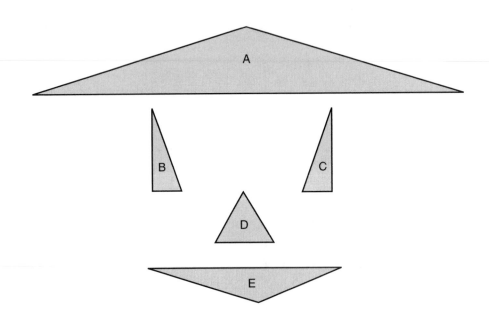

Student pair	Triangle Maker	Isosceles Triangle Maker	Equilateral Triangle Maker	Right Triangle Maker	Obtuse Triangle Maker

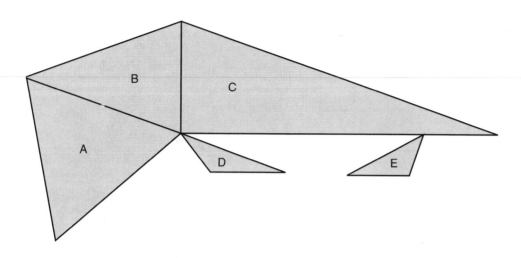

Student pair	Triangle Maker	Isosceles Triangle Maker	Equilateral Triangle Maker	Right Triangle Maker	Obtuse Triangle Maker

TESSELLATION TRANSPARENCY

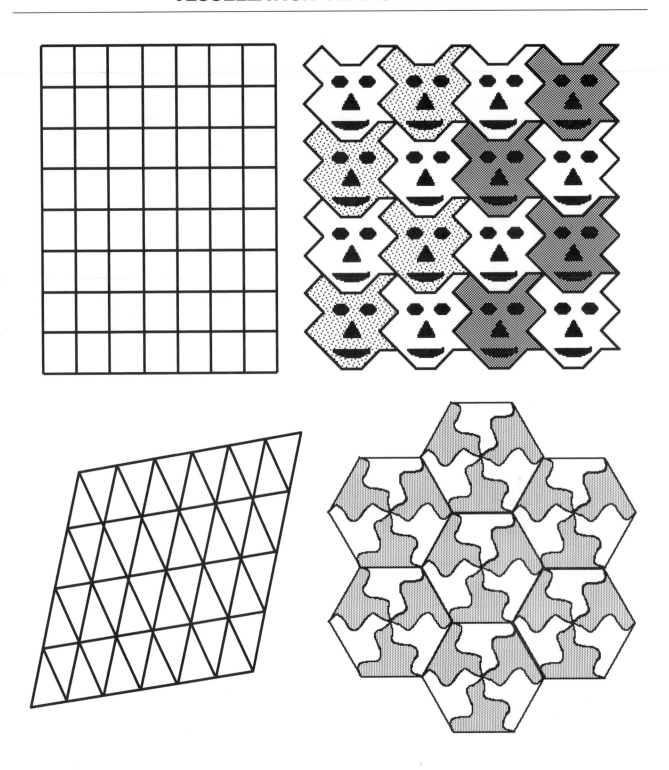